ARROWS, STRUCTURES, AND FUNCTORS

The Categorical Imperative

Academic Press
Rapid Manuscript
Reproduction

ARROWS, STRUCTURES, AND FUNCTORS

The Categorical Imperative

MICHAEL A. ARBIB
Department of Computer and Information Science
University of Massachusetts at Amherst

ERNEST G. MANES
Department of Mathematics
University of Massachusetts at Amherst

ACADEMIC PRESS
A Subsidiary of Harcourt Brace Jovanovich, Publishers
New York London Toronto Sydney San Francisco 1975

ACADEMIC PRESS, INC.
111 Fifth Avenue, New York, New York 10003

United Kingdom Edition published by
ACADEMIC PRESS, INC. (LONDON) LTD.
24/28 Oval Road, London NW1

Library of Congress Cataloging in Publication Data

Arbib, Michael A
Arrows, structures, and functors.

Bibliography: p.
Includes index.
1. Categories (Mathematics) 2. Functor theory.
3. Algebra. I. Manes, Ernest G., joint author.
II. Title.
QA169.A7 512'.55 74-32412
ISBN 0-12-059060-3

PRINTED IN THE UNITED STATES OF AMERICA

83 84 85 9 8 7 6 5 4 3

Pippa
Benjamin
Keira and
Devin

We shall thus have to investigate the possibility of a *categorical* imperative entirely *a priori*, since here we do not enjoy the advantage of having its reality given in experience . . .

There is therefore only a single categorical imperative and it is this: *Act only on that maxim through which you can at the same time will that it should become a universal law*.

Kant arguing, in the *Groundwork of the Metaphysic of Morals*, the need for an introductory treatment, and for the primacy of universal constructions (see page 45 of this volume) as *The Categorical Imperative.*

CONTENTS

PART II

PREFACE

Category theory is the mathematician's attempt to lay bare some of the underlying principles common to diverse fields in the mathematical sciences. It has become, as well, an area of pure mathematics in its own right. Briefly, a *category* is a domain of mathematical discourse characterized in a very general way, and category theory is thus an array of tools for stating results which can be used across a wide mathematical spectrum.

All previous texts on category theory demand considerable mathematical maturity of the reader, assuming familiarity with many different domains of mathematics. As such, the subject matter has been inaccessible to talented undergraduate students, to graduate students in sciences other than pure mathematics, and to professionals in control theory, automata theory, theoretical linguistics, philosophy, mathematical biology and other fields in which category theory is entering the literature. It is for these audiences that the present book is designed.

The contribution of our book, then, is to build up sufficient perspective without demanding more of the reader than a basic knowledge of sets (what is a function? what is a cartesian product of sets?) and matrix theory (what is a linear map? what is a direct product of vector spaces?). In short, this is by far the most elementary category theory text in print and puts advanced books such as Mac Lane's "Categories for the Working Mathematician" within the reach of a much wider audience.

What, then, is *the categorical imperative* – the set of core concepts of category theory which should be shared by this diverse audience before they pursue more specialized avenues tailored to their own area of interest? Our answer is threefold. First is the ability to think with *arrows*: to express key concepts in terms of mappings (we call them *morphisms* in the general setting of category theory) rather than in terms of set elements. Second is the realization that collections of mathematical *structures* find convenient characterization in terms of arrows. Third is the use of *functors* as the appropriate tools with which to compare different domains of mathematical discourse.

The book is divided into two parts. The first part (Chapters 1 to 6) is devoted to arrows and structures: we present the most fundamental aspects of the point of view that a category models a system of 'structures' generalizing specific domains of discourse such as ⟨sets and functions⟩ and ⟨vector spaces and linear maps⟩. The analysis of such systems is in the language of 'arrows' $f: A \longrightarrow B$ whose meaning is that f is a function or linear map respectively in the two examples above. We introduce the reader to the art of 'chasing' *commutative diagrams* as a way of constructing or verifying equalities between chains of arrows. Structures such as monoids, metric spaces and topological spaces are introduced from first principles and the categories induced by these structures are used to demonstrate the diversity of possibilities inherent in generalizing concepts from set theory to category theory. For example, we show that the disjoint union of sets and the free product of groups are both examples of the general categorical notion of a *coproduct*. Chapter 6 gives a categorical perspective on *sets with structure*. A special section applies many of the categorical concepts of Part I to the study of reachability, observability and minimal realization of *automata*. This section should be especially useful to the reader who wishes to go from the basic category theory of this book to the increasing applications in the study, e.g., of computation and control.

The second part of the book (Chapters 7 to 10) is an introduction to *functors*, the third component of our categorical imperative. Just as Euclid's congruence of geometric figures finds modern expression in the concept of a distance-preserving mapping from a geometric space to itself, functors provide the appropriate concept to 'transform' one sort of structure into another – for example, to formalize when two different domains of discourse are 'essentially the same'. Building upon the many examples of Part I we introduce *natural transformations* and *adjoint functors*, starting from the construction of *free* and *cofree* objects. We show that reachability and observability in automata theory correspond to free and cofree constructions respectively. We study *monoidal categories* as a framework for extending set-theoretic constructions to other domains of discourse, noting very briefly that *topos* theory – part of the current research frontier of category theory – offers exciting new vantage points for problems in the foundations of mathematics. Finally, we show that *universal algebra* – another attempt at finding a common language for different branches of mathematics – can itself be captured in category theory language, using the concept of a *monad*. Intriguingly, the language we shall use in bridging universal algebra to category theory will be reminiscent of that used in our categorical treatment of automata theory.

This book grew out of our need to make our work in "Category Theory Applied to Computation and Control" accessible to an audience of computer and system scientists who found existing introductions to category theory overly formidable. We thank Joe Goguen, Stewart Bainbridge, Mitchell Wand, Suad Alagić, Brian Anderson, Hartmut Ehrig and others who have contributed to the growth of this research area; and to the National Science Foundation whose Grant No. GJ 35759 provided support for much of our recent research.

Finally, we express our immense gratitude to Gwyn Mitchell, who typed the master copy from which this book is reproduced. If ever there is an Award for Layout of Commutative Diagrams, we feel sure that it will be hers.

Michael A. Arbib
Ernest G. Manes

Chapter 1

LEARNING TO THINK WITH ARROWS

The reader is familiar with sets as collections of elements, and with functions as assignments of an element in one set to each element of another set. In other words, the usual approach to set theory starts with *elements* and builds all its notions in terms of these. In this chapter, we introduce a different approach to set theory, which builds all its notions in terms of *arrows*, the symbols $f: A \to B$ which represent a function as a unitary whole rather than in element-by-element terms. Then, in subsequent chapters, we shall see that this 'arrow-language' of *category theory* allows us to specify, once and for all, concepts which play an important role in many different areas of mathemetics even though their element-by-element definitions are drastically different in different domains of discourse.

1.1 EPIMORPHISMS AND MONOMORPHISMS

Speaking in terms of elements, we say a function $f: A \longrightarrow B$ (we call A the **domain** of f, and B its **codomain**), which sends each element a of A to some element $f(a)$ of B, is **onto** or **surjective** just in case it maps A onto B in that every b in B is an $f(a)$ for at least one a in A:

$$f: A \longrightarrow B \text{ is onto } \iff B = f(A).$$

For example, the map† $\mathbf{Z} \longrightarrow \mathbf{Z} : n \mapsto n + 1$ of the set $\mathbf{Z}$ of integers into itself is onto; whereas $\mathbf{Z} \longrightarrow \mathbf{Z} : n \mapsto n^2$ is not. For any set B, the **identity map** $\mathrm{id}_B : B \longrightarrow B$ has B as both domain and codomain, and leaves all elements unchanged: $\mathrm{id}_B(b) = b$ for all b in B. Then id_B is certainly onto.

To begin to get the flavor of the categorical approach, we show how to characterize the onto maps without mentioning elements.

Suppose $f: A \longrightarrow B$ is onto, and suppose we have two functions $g: B \longrightarrow C$ and $h: B \longrightarrow C$ with the special property that

$$g(f(a)) = h(f(a)) \text{ for every } a \text{ in } A. \tag{1}$$

† We use **map** or **mapping** as a synonym for 'function'.

Since f is onto, every b in B is $f(a)$ for at least one a in A, and this implies that

$$g(b) = h(b) \quad \text{for every } b \text{ in } B. \tag{2}$$

Now let's get the elements out of (1) and (2). Whenever we have two maps $f : A \longrightarrow B$ and $g : B \longrightarrow C$, such that the codomain of the first equals the domain of the second, we define their **composite** $g{\cdot}f$ to be the map $A \longrightarrow C$: $a \mapsto g(f(a))$ which is defined by the rule "do f, then g". In other words, if we draw the **diagram**

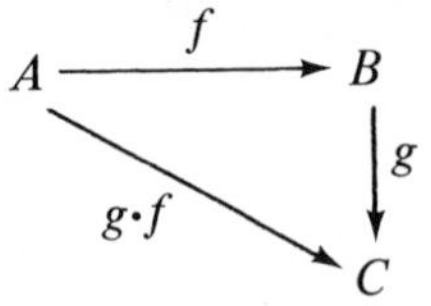

it doesn't matter whether we follow the path $A \xrightarrow{f} B$ followed by $B \xrightarrow{g} C$ to compute $g(f(a))$, or the direct path $A \xrightarrow{g\cdot f} C$ to compute $g{\cdot}f(a)$ – the overall result is the same.

Such diagrams – in which, given a starting point and a destination (such as A and C in the above diagram), different paths yield the same overall function – will occur on almost every page of this book. They are called **commutative diagrams** – *commutare* is the Latin for *exchange*, and we say that a diagram **commutes** if we can exchange paths, between two given points, with impunity.

The other point to make in getting the elements out of (1) and (2) is that two functions are equal just in case they have the same domain, the same co-domain, and yield equal values for equal arguments. Thus we may rephrase (1) as

$$g{\cdot}f = h{\cdot}f \tag{1'}$$

and we may rephrase (2) as

$$g = h. \tag{2'}$$

Summarizing our discussion to date, we have thus proved:

1 If a map $f : A \longrightarrow B$ is onto, then whenever two maps $g, h : B \longrightarrow C$ satisfy $g{\cdot}f = h{\cdot}f$ they must be equal.

This immediately raises the question as to whether the converse is true – let us see that it is by proving that if $f : A \longrightarrow B$ is *not* onto, we can construct two maps $g, h : B \longrightarrow B$ which satisfy $g{\cdot}f = h{\cdot}f$ but which are *not* equal: it then follows that any f for which $g{\cdot}f = h{\cdot}f$ implies $g = h$ must be onto.

To say that $f : A \longrightarrow B$ is not onto is to say that there exists some b_1 in B which is not an $f(a)$ for any a in A. Pick† any $b_2 \neq b_1$, and then define our two

† When is no b_2 available? Can you fix the proof to cover this case?

maps g and h as follows:

$$g = \mathrm{id}_B : B \longrightarrow B$$

$$h(b) = \begin{cases} b \text{ if } b \neq b_1 \\ b_2 \text{ if } b = b_1 \end{cases}$$

Then $g \cdot f(a) = f(a)$ for every a in A, and since $f(a) \neq b_1$ for any a in A, we also have $h \cdot f(a) = f(a)$. Thus $g \cdot f = h \cdot f$ even though it is clear that $g \neq h$, since $g(b_1) \neq h(b_1)$.

What we have now done is shown that an 'element concept' is equivalent to an 'arrow concept' which does not explicitly refer to elements, but only refers to equalities between functions and compositions of function. Let's set out our progress, to date, in terms of a definition and a proposition:

We say a function $f : A \longrightarrow B$ is an **epimorphism** if, for every set C, and for every pair of maps $g : B \longrightarrow C$ and $h : B \longrightarrow C$ such that $g \cdot f = h \cdot f$, we have $g = h$.

Before going further, let's notice that we can abbreviate the definition of epimorphism by taking into account some basic properties of function equality. First, recall that $g \cdot f$ is only defined if the codomain of f equals the domain of g. Secondly, since two equal functions must have equal domains and codomains, to say $g \cdot f = h \cdot f$ implies that codomain g = codomain $g \cdot f$ = codomain $h \cdot f$ = codomain h. Thus, combining these two observations, once we have said "$g \cdot f = h \cdot f$" we have implied that there exists a set C such that both g and h map the codomain of f into C. Thus the definition of an epimorphism can be written simply as:

2 **DEFINITION**: A function f is an **epimorphism** if $g \cdot f = h \cdot f$ always implies $g = h$.

3 **PROPOSITION**: A function $f : A \longrightarrow B$ is onto iff† it is an epimorphism.

□

While we are getting practice at writing things in different ways, we should note that the equality $g \cdot f = h \cdot f$ is often expressed by the mapping diagram

$$A \xrightarrow{f} B \underset{h}{\overset{g}{\rightrightarrows}} C \qquad (3)$$

What we have done here is bend our idea of a commutative diagram a little. With the diagram (3) we say "whether or not $g = h$, it is certainly the case that $g \cdot f = h \cdot f$". We may then say that f is an epimorphism if whenever (3) commutes it follows that $g = h$.

For example, if $f : \mathbf{Z} \longrightarrow \mathbf{Z} : n \mapsto n^2$; $g = \mathrm{id}_{\mathbf{Z}}$; while $h : \mathbf{Z} \longrightarrow \mathbf{Z}$ sends n

---

† "iff" is an abbreviation for "if and only if".

to n if $n \neq 3$, but satisfies $h(3) = 4$, we have that (3) commutes, even though $g \neq h$.

Our convention for commutative diagrams is then that any two paths between two given points must yield the same overall function *so long as at least one path involves more than one arrow*. Thus if the diagram

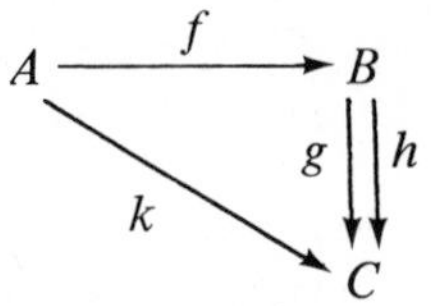

commutes then $g \cdot f = h \cdot f = k$, but we do not know whether or not $h = g$.

Having gained some feel for arrows and diagrams, we can now 'remove the elements' from one-to-one maps with relative ease as follows. A function $f : B \longrightarrow A$ is **one-to-one** or **injective** just in case $f(b_1) \neq f(b_2)$ in A whenever $b_1 \neq b_2$ in B.

Now suppose two maps $g, h : C \longrightarrow B$ have the special property that

$$f(g(c)) = f(h(c)) \quad \text{for every } c \text{ in } C.$$

By the definition of one-to-one, this must imply that

$$g(c) = h(c) \quad \text{for every } c \text{ in } C.$$

In other words,

4 If a map $f : B \longrightarrow A$ is one-to-one, then whenever two maps $g, h : C \longrightarrow B$ satisfy $f \cdot g = f \cdot h$, they must be equal.

This is a dramatic development, for while the definitions of one-to-one and onto look very different, the previous sentence is 'the mirror image of' 1. Think about this for yourself now, and we shall discuss it explicitly below.

We now verify the converse of 4 by showing that if $f : B \longrightarrow A$ is not one-to-one, we can construct $g, h : C \longrightarrow B$ for which $f \cdot g = f \cdot h$ even though $g \neq h$:

Suppose f is not one-to-one, so that we can find distinct elements b_1 and b_2 in B for which $f(b_1) = f(b_2)$. Then take g to be id_B, while $h : B \longrightarrow B$ is defined by

$$h(b) = \begin{cases} b & \text{if } b \neq b_1 \\ b_2 & \text{if } b = b_1 \end{cases}$$

(which should look familiar!). Clearly $f \cdot g(b) = f(b)$ for every b in B, while $f \cdot h(b) = f(b)$ for $b \neq b_1$, and $f \cdot h(b_1) = f(b_2) = f(b_1)$ by our choice of b_1 and b_2. Thus $f \cdot g = f \cdot h$ while $g \neq h$.

Again, we summarize our progress in a definition and a proposition:

5 DEFINITION: A function f is a **monomorphism** if $f \cdot g = f \cdot h$ always implies $g = h$.

6 **PROPOSITION**: A function f is one-to-one iff it is a monomorphism. □

Now we can return to the similarity between 1 and 4 by contrasting the diagrams for definitions 2 and 5:

7 f is an epimorphism if $A \xrightarrow{f} B \overset{g}{\underset{h}{\rightrightarrows}} C$ implies $g = h$.

8 f is a monomorphism if $C \overset{g}{\underset{h}{\rightrightarrows}} B \xrightarrow{f} A$ implies $g = h$.

7 and 8 can each be obtained from the other by 'reversing the arrows' as becomes clear if we rewrite 8 in the form

9 f is a monomorphism if $A \xleftarrow{f} B \overset{g}{\underset{h}{\leftleftarrows}} C$ implies $g = h$.

This is our first taste of an important idea that recurs throughout category theory: *important ideas come in pairs.* Each member of the pair is called the **dual** of the other, and its definition is obtained from the definition of its dual by reversing all the arrows.

We use **category** to denote a domain of discourse. So far, our domain of discourse has been the category **Set** of all sets and mappings. To give some foretaste of the abstract discussion of Chapter 2, and to give more meat to our discussion of duality, we introduce a new domain of discourse: the category $\mathbf{Set}^{\mathrm{op}}$ is the **opposite**, or mirror, category of **Set**. It still uses sets as its basic objects, but instead of having arrows $\longrightarrow$ which denote the operation of some function, it has arrows $\longrightarrow\!\!<$ which indicate a **reversed** presentation of a function. In other words, when we say $A \overset{f}{\longrightarrow\!\!<} B$ is an arrow in $\mathbf{Set}^{\mathrm{op}}$, we do *not* say f is a function from A to B: we are simply saying that f is the label of some function in the *opposite* direction, $f : B \to A$ in **Set**. We say $A_1 \overset{f_1}{\longrightarrow\!\!<} B_1$ equals $A_2 \overset{f_2}{\longrightarrow\!\!<} B_2$ in $\mathbf{Set}^{\mathrm{op}}$ iff $B_1 \xrightarrow{f_1} A_1$ equals $B_2 \xrightarrow{f_2} A_2$ in **Set**, i.e. iff $A_1 = A_2$, $B_1 = B_2$, and $f_1(b) = f_2(b)$ for every b in B_1. Again, composition in $\mathbf{Set}^{\mathrm{op}}$ is given by

$$A \overset{f}{\longrightarrow\!\!<} B \overset{g}{\longrightarrow\!\!<} C \text{ is the opposite of } C \xrightarrow{g} B \xrightarrow{f} A$$

so that $(g \cdot f)$ in $\mathbf{Set}^{\mathrm{op}}$ is the opposite of $(f \cdot g)$ in **Set**.

Now let **K** denote either **Set** or $\mathbf{Set}^{\mathrm{op}}$, and let us rephrase definitions 2 and 5 so that they apply to $\mathbf{Set}^{\mathrm{op}}$ as well as to **Set**:

10 An arrow f in **K** is an **epimorphism** in **K** if the equality $g \cdot f = h \cdot f$ in **K** always implies $g = h$.

11 An arrow f in **K** is a **monomorphism** in **K** if the equality $f \cdot g = f \cdot h$ in **K** always implies $g = h$.

Now reversing the arrows in $\mathbf{Set}^{\mathrm{op}}$ just takes us back to **Set**, and so we have $\mathbf{K}^{\mathrm{op}} = \mathbf{Set}$ if $\mathbf{K} = \mathbf{Set}^{\mathrm{op}}$ while, of course, $\mathbf{K}^{\mathrm{op}} = \mathbf{Set}^{\mathrm{op}}$ when $\mathbf{K} = \mathbf{Set}$. Recalling that the order of composition is reversed when we pass from **K** to $\mathbf{K}^{op}$ we then have:

12 OBSERVATION: f is a monomorphism in **K** iff f is an epimorphism in $\mathbf{K}^{op}$. □

In other words, we say that W (whatever that may prove to be) is an example of the **dual** of W' just in case each instance of W in **K** is an instance of W' in $\mathbf{K}^{op}$: monomorphism is dual to epimorphism because every f which is a monomorphism in **K** is an epimorphism in $\mathbf{K}^{op}$.

The preceding (starting at 7) has been rather heady stuff, and may seem very strange at the first reading or two. Fortunately, little else in this chapter is at such a high level of abstraction, so that you may read on even if our discussion of $\mathbf{Set}^{op}$ has mystified you. However, we suggest that each time duality is mentioned in the remainder of this chapter, you reread the discussion of $\mathbf{Set}^{op}$. Hopefully, by the time you come to our general discussion of opposite categories in Chapter 2, the example of $\mathbf{Set}^{op}$ will have come to seem very natural to you.

To close this section we study the bijections: maps which are both onto and one-to-one. We may thus think of a bijection $f: A \longrightarrow B$ as being a *relabelling* – each b in B serves to label one and only one a in A, namely that for which $b = f(a)$. That f is onto says that at least one such a exists; that f is one-to-one says that the a is unique.

If $A = \{1, 2, 3\}$ and $B = \{3, 7, 11\}$ then the f defined by

$$f(1) = 3; \quad f(2) = 11; \quad f(3) = 7$$

is a bijection. Again, if $A = \mathbf{Z}$ while B is the set of all *even* integers, then

$$f: A \longrightarrow B : n \mapsto 2n$$

is also a bijection.

If f is a bijection, we can clearly define a function $k: B \longrightarrow A$ by the rule

$$k(b) = a \quad \text{iff} \quad f(a) = b$$

and we then have that

$$f(k(b)) = b \quad \text{for every } b \text{ in } B\text{; and}$$

$$k(f(a)) = a \quad \text{for every } a \text{ in } A$$

which in our succinct vocabulary of composition and identity functions may be reexpressed as

$$f \cdot k = \mathrm{id}_B \quad \text{and} \quad k \cdot f = \mathrm{id}_A .$$

This suggests the following:

13 DEFINITION: A function $f: A \longrightarrow B$ is an **isomorphism** iff there exists a function $k: B \longrightarrow A$ which satisfies

$$f \cdot k = \mathrm{id}_B \quad \text{and} \quad k \cdot f = \mathrm{id}_A .$$

We call k the **inverse** of f. Moreover, we say two sets A and B are **isomorphic** just in case there exists at least one function $f : A \longrightarrow B$ such that f is an isomorphism, and write $A \cong B$.

Note that the inverse k is unique: if k' is also an inverse of f, we have that $f \cdot k'$ also equals id_B. But then

$$k = k \cdot \mathrm{id}_B = k \cdot (f \cdot k') = (k \cdot f) \cdot k' = \mathrm{id}_A \cdot k' = k'.$$

As the reader may well suspect, we can easily prove:

14 PROPOSITION: f is a bijection iff f is an isomorphism.

Proof: We have already proved that each bijection is an isomorphism. We now give two proofs of the converse:
'Old-fashioned proof': We manipulate elements to prove that the isomorphism $f : A \longrightarrow B$ is one-to-one and onto. Let k be the inverse of f.

To see that f is onto, pick any b in B, and note that $f(k(b)) = \mathrm{id}_B(b) = b$.

To see that f is one-to-one, suppose $f(a_1) = f(a_2)$. Then $k(f(a_1)) = \mathrm{id}_A(a_1) = a_1$, while $k(f(a_2)) = a_2$, so that $a_1 = a_2$.
'Modern Proof': We manipulate diagrams to prove that the isomorphism $f : A \longrightarrow B$ is an epimorphism and a monomorphism. Let k be the inverse of f.

To see that f is an epimorphism, note that

$$g \cdot f = h \cdot f \Rightarrow g \cdot f \cdot k = h \cdot f \cdot k \Rightarrow g = h.$$

To see that f is a monomorphism, note that if

$$f \cdot g = f \cdot h \quad \text{then} \quad k \cdot f \cdot g = k \cdot f \cdot h \quad \text{and then} \quad g = h. \; \square$$

Let's reexpress the definition of an isomorphism in a commutative diagram:

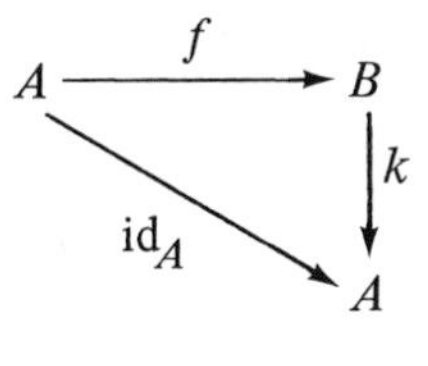

says $k \cdot f = \mathrm{id}_A$

while

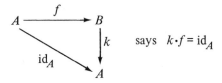

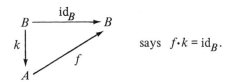

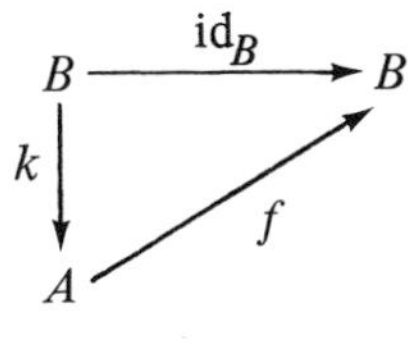

says $f \cdot k = \mathrm{id}_B$.

Equivalently and more succinctly:

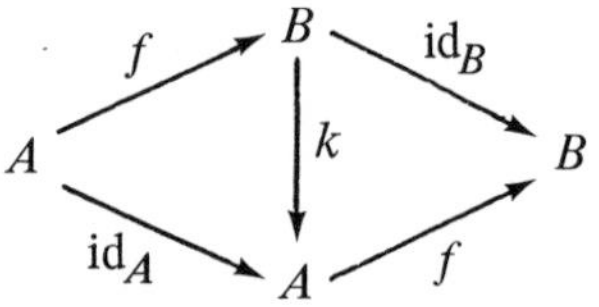

What happens if we reverse the arrows to get the diagram in $\mathbf{Set}^{\text{op}}$? We get:

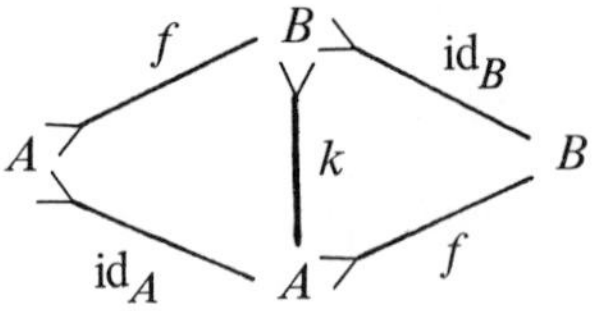

which says that, in $\mathbf{Set}^{\text{op}}$, we have $f \cdot k = \text{id}_A$ and $k \cdot f = \text{id}_B$ so that f *is an isomorphism with inverse* k *in* $\mathbf{Set}^{\text{op}}$ *as well as in* **Set**. Thus *isomorphism is a self-dual concept.* We can see this another way: an isomorphism is an epimorphism and a monomorphism. Thus the dual of an isomorphism is both the dual of an epimorphism and the dual of a monomorphism; i.e. it is both a monomorphism and an epimorphism; i.e. it is itself an isomorphism.

We close this section by noting that every map $f : A \longrightarrow B$ of one set into another has the factorization

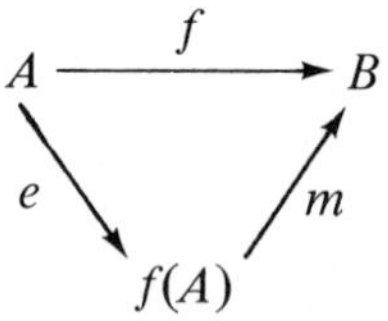

where $e : A \longrightarrow f(A)$ is the epimorphism $a \mapsto f(a)$, and where m is the monomorphism $f(A) \longrightarrow B : b \mapsto b$. Clearly $f(a) = m \cdot e(a)$. We now point out that this factorization is 'unique up to isomorphism':

15 DEFINITION: We say that the pair (e, m) is an **epi-mono factorization** of the map $f : A \longrightarrow B$ if e is an epimorphism and m is a monomorphism such that $f = m \cdot e$.

16 PROPOSITION: Epi-mono factorizations are *unique up to isomorphism* in the sense that if (e', m') is an epi-mono factorization

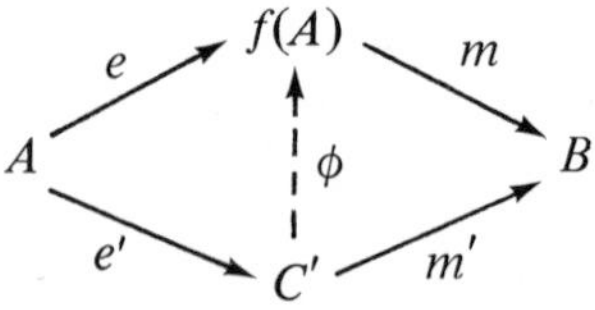

of a given map $f : A \longrightarrow B$, then there is an isomorphism ϕ such that the above diagram commutes.

Proof: This result needs an element-by-element proof. The 'arrow-proof' requires a stronger arrow-definition of onto maps than that provided by epimorphisms – namely the coequalizers of Section 1.3.

Let $c' \in C'$. Pick $a \in A$ with $e'(a) = c'$. Define $\phi(c') = e(a)$. Let $a \in A$. Then there exists $a_1 \in A$ with $e'(a) = e'(a_1)$ and $\phi(e'(a)) = e(a_1)$. But $m(e(a)) = f(a) = m'(e'(a)) = m'(e'(a_1)) = m(e(a_1))$ and m is one-to-one, so that $e(a) = e(a_1) = \phi(e'(a))$. This proves that $\phi{\cdot}e' = e$. It is then immediate that ϕ is onto since if $a \in A$ (so that $e(a)$ is a typical element of $f(A)$), $e(a) = \phi(c')$ for $c' = e'(a)$. Let $c' \in C'$. Then $\phi(c) = e(a)$ where $e'(a) = c'$ and so $m(\phi(c')) = m(e(a)) = m'(e'(a)) = m(c')$. This proves $m{\cdot}\phi = m'$. Finally, if $c' \neq c'' \in C'$, then since $m'(c') \neq m'(c'')$, $m(\phi(c')) \neq m(\phi(c''))$ so that surely $\phi(c') \neq \phi(c'')$. □

Exercises

1 Prove that if $f : A \longrightarrow B$ is one-to-one then f is a *split monomorphism*, i.e. that there exists $g : B \longrightarrow A$ with $g{\cdot}f = \mathrm{id}_A$.

2 Prove that for every function $f : A \longrightarrow B$ with A non-empty, there exists a 'choice function' $g : B \longrightarrow A$ such that $f{\cdot}g{\cdot}f = f$. In axiomatic set theory this is an acceptable form of the 'axiom of choice'.

3 Let A be any set and let $f : \emptyset \longrightarrow A$ be the unique function 'inclusion of the empty set'. [Note: recall that a function $A \longrightarrow B$ is any subset f of $A \times B$ satisfying "given $a \in A$ there exists a unique pair in f with first coordinate a.] Prove that f is a monomorphism.

4 Let 2^A denote the *power set* of all subsets of A. Given $f : A \longrightarrow B$ define $2^f : 2^B \longrightarrow 2^A$ by $2^f(S) = f^{-1}(S) = \{a \in A \mid f(a) \in B\}$. Prove that f is onto iff 2^f is one-to-one.

5 Using the definitions of exercise 4, prove that if (e, m) is an epi-mono factorization of f then $(2^m, 2^e)$ is an epi-mono factorization of 2^f.

1.2 PRODUCTS AND COPRODUCTS

The starting point of analytic geometry (also known as coordinate geometry, or as cartesian geometry, in honor of its founder Réné Descartes) is the observation that every point of the plane can be uniquely represented by the pair (x, y) of its coordinates (Figure 1).

Generalizing this notion to arbitrary sets, we call the set of all ordered pairs with first element from A and second element from B,

$$A \times B = \{(a, b) \mid a \in A \text{ and } b \in B\},$$

the **product** (also known as the **cartesian product**) of A and B, in that order. As

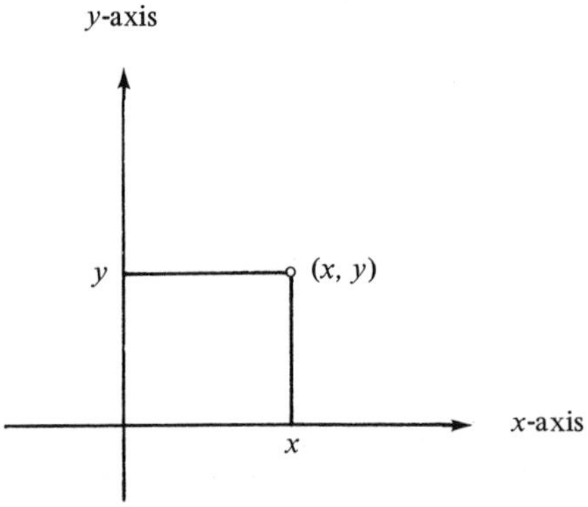

Figure 1 The Cartesian Plane.

a test of our concepts from Section 1, note that the map $f: A \times B \longrightarrow B \times A$: $(a, b) \mapsto (b, a)$ has for inverse $k: B \times A \longrightarrow A \times B : (b, a) \mapsto (a, b)$, so that $A \times B \cong B \times A$. We see that Figure 1 represents the plane as $\mathbf{R} \times \mathbf{R}$, the product of two copies of the set $\mathbf{R}$ of real numbers.

By now, the reader should be asking, "Can we characterize $A \times B$ in terms of arrows, rather than element-by-element in terms of ordered pairs?" The answer is, once again, "yes," but – as in our discussion of onto and one-to-one maps – the arrow-definition will look very different from the element-definition.

Our first step is to note that there are two very special maps associated with $A \times B$:

$$\pi_1 : A \times B \longrightarrow A\ ,\ (a, b) \mapsto a$$

and

$$\pi_2 : A \times B \longrightarrow B\ ,\ (a, b) \mapsto b.$$

We call these maps **projections**, motivated by the example of Figure 1 in which each of these maps projects a point down upon one of the axes.

Suppose, now, we are given any other set C and any two maps $p_1 : C \longrightarrow A$ and $p_2 : C \longrightarrow B$. In this case we may design a new map $p : C \longrightarrow A \times B$ by the rule $p(c) = (p_1(c), p_2(c))$. Thus we have that p makes the following diagram commute:

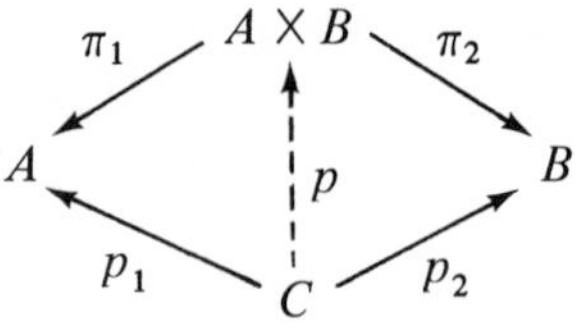

(1)

where the solid arrows indicate the maps we were given to start with, while the broken arrow is to be 'filled in' in such a way as to make the whole diagram commute. In fact, the p constructed is the only one possible, for if $p(c) = (a, b)$, then commutativity implies

$$a = \pi_1 \cdot p(c) = p_1(c)$$
$$b = \pi_2 \cdot p(c) = p_2(c).$$

Thus we have the following property:

1 The pair of maps $\{A \times B \xrightarrow{\pi_1} A,\ A \times B \xrightarrow{\pi_2} B\}$ is such that, given any pair of maps $\{C \xrightarrow{p_1} A,\ C \xrightarrow{p_2} B\}$ there is a unique map $C \xrightarrow{p} A \times B$ for which the diagram (1) commutes.

Let's turn this statement into a formal definition, in just the way that we obtained the definitions of epimorphism and monomorphism in Section 1:

2 **DEFINITION**: A **product** of two sets A_1 and A_2 is a set A equipped with two maps $\pi_1 : A \longrightarrow A_1$ and $\pi_2 : A \longrightarrow A_2$ (called **projections**) with the property that, given any other set C with pair of maps $p_1 : C \longrightarrow A_1$ and $p_2 : C \longrightarrow A_2$, there exists a unique map p such that†

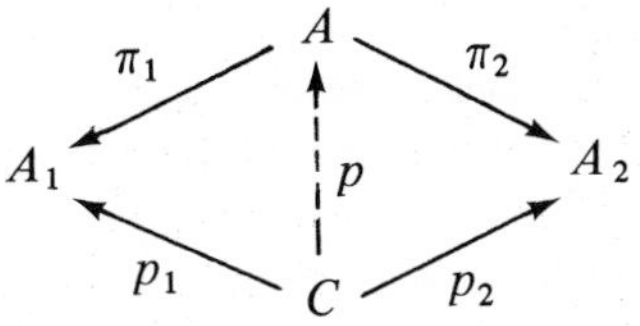

Note that we may refer to the product as being the ordered pair $(\pi_1 : A \longrightarrow A_1,\ \pi_2 : A \longrightarrow A_2)$, since the set A is implicit, being the domain of both π_1 and π_2.

Just as we proved, in Section 1, that every monomorphism was a one-to-one map, the reader might expect that we can now prove that the only product of A_1 and A_2 must be the cartesian product $A_1 \times A_2$. But this is false! If we take A, in definition 2, to be $A_2 \times A_1$ and then take $\pi_1(a_2, a_1) = a_1$ and $\pi_2(a_2, a_1) = a_2$, the diagram

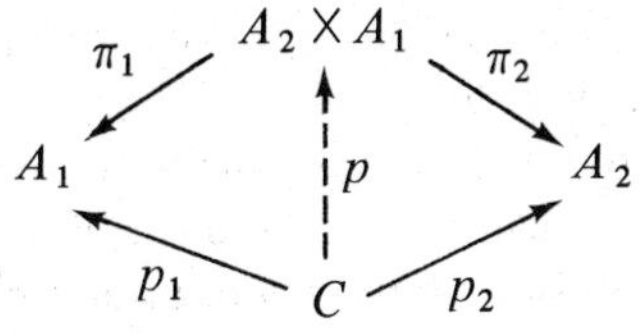

defines a unique map p by the rule $p(c) = (p_2(c), p_1(c))$.

Does this mean that our attempt to characterize $A_1 \times A_2$ with arrows lies in ruins? Not really, for we have already noted that $A_2 \times A_1$ is *isomorphic* to $A_1 \times A_2$ – in other words, the two sets differ only by the relabelling of elements

---

† Henceforth we shall often use "such that[diagram]" as an abbreviation for "such that the diagram [diagram] commutes".

which rewrites (a_1, a_2) as (a_2, a_1). It should be clear, then, that once a set is a product, relabelling its elements will not change that fact. The most we can hope for, then, is that every product of A_1 and A_2 is isomorphic to $A_1 \times A_2$. This proves to be the case:

3 **PROPOSITION**: If $(\pi_1 : A \to A_1,\ \pi_2 : A \to A_2)$ and $(\pi_1' : A' \to A_1,\ \pi_2' : A' \to A_2)$ are both products of A_1 and A_2, then they are isomorphic, i.e. there exists an isomorphism $\phi : A' \to A$ such that $\pi_1' = \pi_1 \cdot \phi$ and $\pi_2' = \pi_2 \cdot \phi$. In other words, the product of two sets is *unique up to isomorphism.*

Proof: We give a genuine arrow-chasing proof without any mention of elements: Consider the diagram

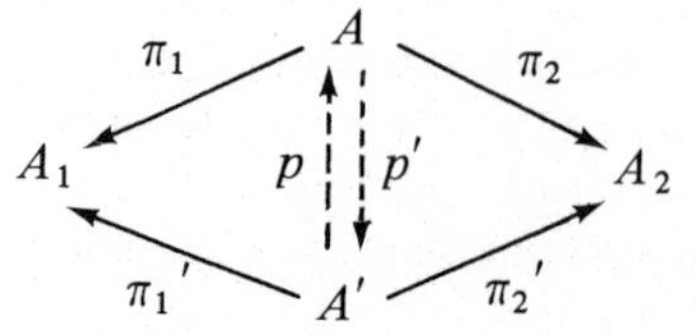

Since A is a product, we can find a unique p such that $\pi_1 \cdot p = \pi_1'$ and $\pi_2 \cdot p = \pi_2'$. Since A' is a product (if you look at the diagram 'upside-down', it is just the product diagram for A': the way a diagram faces on the page is irrelevant to the information it conveys) we can find a unique p' such that $\pi_i' \cdot p' = \pi_i$ for $i = 1, 2$. Now consider the diagram

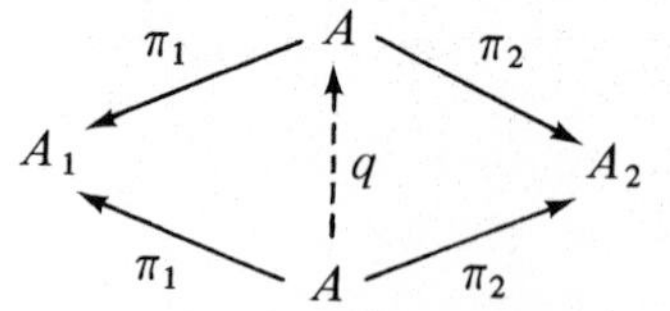

Since A is a product, there is a unique q which will make the diagram commute, and clearly $q = \mathrm{id}_A$ will do the job. But the equalities $\pi_i \cdot p = \pi_i'$ and $\pi_i' \cdot p' = \pi_i$ tell us that $\pi_i \cdot p \cdot p' = \pi_i$, so that $q = p \cdot p'$ will also do the job. But we said that q was unique, and so $p \cdot p' = \mathrm{id}_A$.

The same argument but with A' replacing A in the last diagram, tells us that $p' \cdot p = \mathrm{id}_{A'}$. But the two equalities $p \cdot p' = \mathrm{id}_A$ and $p' \cdot p = \mathrm{id}_{A'}$ are precisely the conditions that p be an isomorphism with inverse p'. Thus p is the desired ϕ of our theorem. □

4 **COROLLARY**: $(\pi_1 : A \to A_1,\ \pi_2 : A \to A_2)$ is a product of A_1 and A_2 iff $A \cong A_1 \times A_2$ by an isomorphism transforming π_i into the map $(a_1, a_2) \mapsto a_i$. □

Because of this result, we shall often denote *any* A which satisfies definition 2 by $A_1 \times A_2$ whether or not its elements are actually labelled (a_1, a_2) with a_1

in A_1 and a_2 in A_2.

With this study of products, we have laid bare another general aspect of the methodology of category theory: *we are not interested in the differences between two objects, or collections of functions, if they are isomorphic.* In this case, we consider two sets as the same if they only differ by relabelling, and hence we regard a definition, such as 2, as defining a set uniquely if it defines a set uniquely up to an isomorphism; i.e. if whenever A and A' both satisfy the criteria of the definition, we must have $A \cong A'$.

At this stage, we may recall another general principle, adumbrated in Section 1, namely that *important ideas come in pairs,* and so ask: what is the *dual* of a product, i.e. what concept is defined by reversing the arrows in the definition of a product? It is customary to call the dual of W a co-W, save where, as in the case of monomorphism and epimorphism, each member of the dual pair has a well-established name of its own. Reversing the arrows in 2 then yields the

5 **DEFINITION**: A **coproduct** of two sets A_1 and A_2 is a set A equipped with two maps $\mathrm{in}_1 : A_1 \longrightarrow A$ and $\mathrm{in}_2 : A_2 \longrightarrow A$ (called **injections**) with the property that, given any other set C with a pair of maps $q_1 : A_1 \longrightarrow C$ and $q_2 : A_2 \longrightarrow C$, there exists a unique map q such that

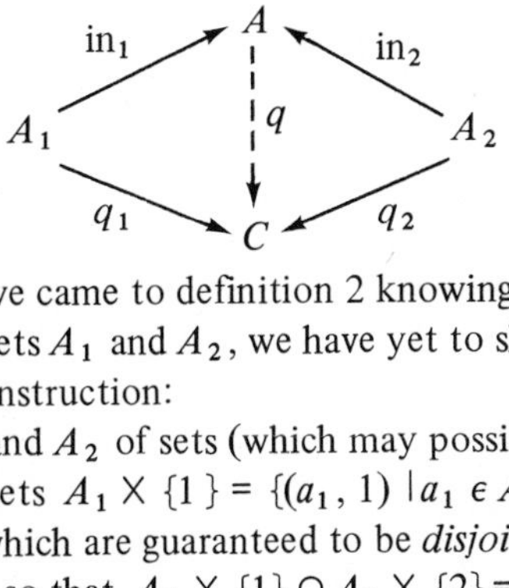

However, whereas we came to definition 2 knowing that a product $A_1 \times A_2$ existed for any pair of sets A_1 and A_2, we have yet to show that each pair has a coproduct. Here is a construction:

Given any pair A_1 and A_2 of sets (which may possibly have elements in common) we form the sets $A_1 \times \{1\} = \{(a_1, 1) \mid a_1 \in A_1\}$ and $A_2 \times \{2\} = \{(a_2, 2) \mid a_2 \in A_2\}$ which are guaranteed to be *disjoint*, since whenever $(a, k) = (a', k')$, $k = k'$, so that $A_1 \times \{1\} \cap A_2 \times \{2\} = \emptyset$. We then let $A_1 + A_2$ be the union $A_1 \times \{1\} \cup A_2 \times \{2\}$ – we call it the **disjoint union** of A_1 and A_2 because we ensured that the two sets were disjoint before uniting them – and define the maps

$$\mathrm{in}_k : A_k \longrightarrow A_1 + A_2 : a \mapsto (a, k) \quad \text{for } k = 1, 2$$

which simply sends each element to its relabelled version in $A_1 + A_2$. Then, given any diagram of the form

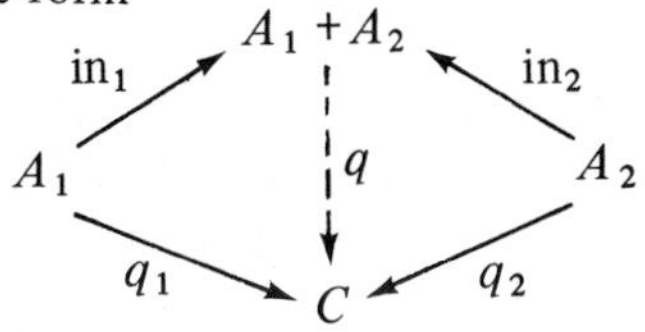

we see that there is a unique q which completes it, namely

$$q(a_k, k) = q_k(a_k).$$

Following 2 and 3 we then have:

6 **PROPOSITION:** The coproduct of two sets is unique up to isomorphism.

Proof: The reader should carefully review the discussion of $\mathbf{Set}^{\mathrm{op}}$ following 1.9, and note that the following proof is acceptable: "A coproduct in **Set** is a product in $\mathbf{Set}^{\mathrm{op}}$. The proof of 3 uses arrows, not elements, and so is valid in $\mathbf{Set}^{\mathrm{op}}$ as well as in **Set**. Thus products are unique in $\mathbf{Set}^{\mathrm{op}}$, and so coproducts are unique up to isomorphism in **Set**." □

If this proof smacks too much of black magic to the reader at this early stage of her initiation into the lore of arrow-thought, we suggest that she write out the proof of 3 with arrows reversed, to check that it does yield a valid proof of 6. Then note that the proof we gave above of 6 is just a formal way of saying, "Writing out the proof for products with arrows reversed will prove this new result."

In fact, our proof of 6 can be subsumed under the following general rule:

If a W has been proved 'unique up to isomorphism' in **K** *using arrows, not elements, then a W is 'unique up to isomorphism' in* $\mathbf{K}^{\mathrm{op}}$ *as well as in* **K**. *Thus co-W's are also 'unique up to isomorphism' in* **K**.

Because of this, when we state a result of the form "Co-W's have property co-P" when we have already used arrows to prove that "W's have property P", we may prove it simply by tersely writing *by duality*, or may not offer a proof at all. In fact, as we grow more experienced, we may not even bother to note the fact that "Co-W's are co-P", but simply use it any time after our proof that "W's are P".

7 **COROLLARY:** $(\mathrm{in}_1 : A_1 \longrightarrow A,\ \mathrm{in}_2 : A_2 \longrightarrow A)$ is a coproduct of A_1 and A_2 iff $A \cong A_1 + A_2$ by an isomorphism transforming in_k into the map $a \mapsto (a, k)$. □

To close this section, we note that we can form products of arbitrary collections of sets, even infinite collections. We leave the corresponding generalization for coproducts as exercise 3. First, however, we show that we may define $A_1 \times A_2$ not as a set of ordered pairs but as a set of functions:

$$A_1 \times A_2 = \{f \mid f : \{1, 2\} \longrightarrow A_1 \cup A_2 \ \text{ with } f(1) \in A_1 \text{ and } f(2) \in A_2\}. \quad (2)$$

To see this, we need simply note that the correspondences

$$(a_1, a_2) \mapsto \text{the } f \text{ with } f(1) = a_1 \text{ and } f(2) = a_2$$

$$f \mapsto (f(1), f(2))$$

give the isomorphism and its inverse which establish that (2), together with the projections $f \mapsto f(1)$ and $f(2)$, is a product of A_1 and A_2. But (2) immediately generalizes to any family $\{A_i \mid i \in I\}$ of sets, with one set A for each index i in the (possibly infinite) index set I:

8 The **cartesian product** of the family $(A_i \mid i \in I)$ of sets is the set

$$\prod_{i \in I} A_i = \{f \mid f : I \longrightarrow \bigcup_{i \in I} A_i \text{ with } f(i) \in A_i \text{ for each } i \in I\}$$

together with the projections $\pi_j : \prod_{i \in I} A_i \longrightarrow A_j,\ f \mapsto f(j)$.

Generalizing 2 we then have

9 **DEFINITION:** A **product** of the family of sets $(A_i \mid i \in I)$ is a set A equipped with a family of maps $(\pi_i : A \longrightarrow A_i \mid i \in I)$ (called **projections**) with the property that, given any other set C with a family of maps $(p_i : C \longrightarrow A_i \mid i \in I)$ there exists a unique map p such that

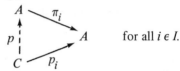

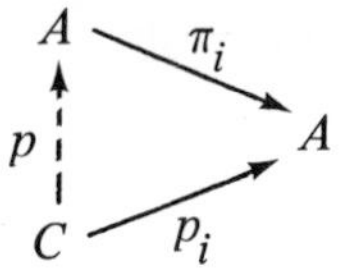

for all $i \in I$.

To make sure that the reader has really comprehended the story so far we ask the reader to work the exercises which conclude this section.

Exercises

1 Prove that the cartesian product of 8 is a product in the sense of 9.

2 Prove that every product in the sense of 9 is isomorphic to the cartesian product of 8. [Hint: The proof of 3 requires very little rewriting to go through in the general case.]

3 Generalize the definition of the disjoint union A_1 and A_2 of two sets to define the disjoint union $\coprod A_i$ of an arbitrary family $(A_i \mid i \in I)$ of sets. [Hint: Define $\coprod A_i = \{(a, i) \mid a \in A_i\}$.] Generalize 5 to give the definition of coproduct of $(A_i \mid i \in I)$. Prove that each such coproduct is isomorphic to $\coprod_{i \in I} A_i$. [Hint: Duality!]

4 Verify that if I has only one element, 1, then

$$\prod_{i \in I} A_i \cong A_1 \cong \coprod_{i \in I} A_i.$$

5 The interesting thing about general definitions is that they are often more general than you realize. Consider what happens to $\prod_{i \in I} A_i$ and $\coprod_{i \in I} A_i$ when $I = \emptyset$, the empty set.

$$\prod A_i = \{f \mid f : \emptyset \longrightarrow \emptyset\} \quad \text{and so has exactly one element.}$$

What is the empty coproduct?

6 In the context of 8, let $h : \amalg A_i \longrightarrow I$ be defined by $h(a, i) = i$. Prove that ΠA_i may be identified with the set of all choice functions for h in the sense of exercise 1.2.

7 For each set A, the set A^* of all 'words on the alphabet A' may be defined as

$$A^* = \coprod_{n=0}^{\infty} A^n$$

(where $A^0 = \{\Lambda\}$ 'the empty word', $A^{n+1} = A^n \times A$). Why can't we use an ordinary union? [Hint: Construct a set A such that $A \cap (A \times A) \neq \emptyset$.]

8 Prove that every set is isomorphic to a coproduct of 1-element sets.

9 Prove that the set $A \times B$ is isomorphic to a coproduct of B copies of A. [Hint: In figure 1, think of the plane as a union of horizontal lines.]

3 COEQUALIZERS AND EQUALIZERS

The notion of equivalence relation is one of the most fundamental in modern mathematics, and yet proves strangely troubling to those who have not yet attained 'mathematical maturity'. Because of this, we use the idea of 'barrels' to get the notion across:

Suppose we have a set A of apples and a set B of barrels, and we sort the apples into the barrels in such a way that there is at least one apple in each barrel. In this way we have defined an onto map (since our goal now is to build an alternative arrow-formulation for onto maps, we will *not* refer to onto maps as epimorphisms at this stage)

$$s : A \longrightarrow B, \quad a \mapsto \text{the barrel in which } a \text{ is sorted.}$$

The language of functions, then, specifies an onto map which tells of each apple into which barrel it is sorted. The language of equivalence relations, by contrast, tells us, of any apple, which other apples are sorted into the same barrel. In other words, sorting defines the **relation S on A** (i.e. the subset S of $A \times A$),

$$S = \{(a_1, a_2) \in A \times A \mid a_1 \text{ and } a_2 \text{ are sorted into the same barrel}\}.$$

Now this relation S has three special properties, each of which has its own special name:

Reflexivity: $(a, a) \in S$ for every a in A. [Every apple is in the same barrel as itself.]

Symmetry: $(a, a') \in S$ iff $(a', a) \in S$ for all a, a' in A. [a and a' are in the same barrel iff a' and a are in the same barrel.]

Transitivity: If $(a, a') \in S$ and $(a', a'') \in S$ then $(a, a'') \in S$ for all a, a' and a'' in A. [If a is in the same barrel as a', which is in the same barrel as a'', then a and a'' are in the same barrel.]

Can we recapture B from S? For each a in A, let us use $[a]$ to denote the set of apples (including a) which lie in the same barrel as a:

$$[a] = \{a' \in A \mid (a, a') \in S\}.$$

We call $[a]$ the **equivalence class of** a **with respect to** A; and write A/S for the **quotient set** whose elements are the distinct equivalence classes. What we shall see is that we may think of a barrel simply as a barrel, or we may identify it with the set of apples it contains. Thus A/S, which is a set of sets, may be thought of quite concretely as a set of barrels as follows.

If a and a' are distinct apples lying in the same barrel, we have $[a] = [a']$, while if a and a' lie in distinct barrels, it is clear that $[a] \neq [a']$. Thus, we have defined an isomorphism (relabelling)

$$[a] \leftrightarrow s(a)$$

which we run backwards by, given a barrel b, picking an apple a from it and then taking the set of apples which share the barrel with a:

$$b \leftrightarrow [a] \qquad \text{for any apple } a \text{ in } b$$

simply relabels a barrel by the set of apples sorted into it: $B \cong S/A$. It is clear, too, that the map $A \longrightarrow S/A : a \leftrightarrow [a]$ is 'really' the onto sorting map s, apart from this relabelling.

With this motivation available, we may now give a swift formal account of the relation between equivalence relations and onto maps.

1 **DEFINITION**: An **equivalence relation** E on a set A is a relation on A which satisfies the three conditions:

Reflexivity: $(a, a) \in E$
Symmetry: $(a, b) \in E$ iff $(b, a) \in E$
Transitivity: If $(a, b) \in E$ and $(b, c) \in E$ then $(a, c) \in E$ for all a, b, c in A.

2 **DEFINITION**: Given an equivalence relation E on a set A, we define the **equivalence class** of an element a of A with respect to E to be

$$[a]_E = \{a' \mid a' \in A \text{ and } (a, a') \in E\}.$$

(When no ambiguity can arise, we write $[a]$ for $[a]_E$.) The **factor set** or **quotient set of** A **with respect to** E is then the set of equivalence classes

$$A/E = \{[a] \mid a \in A\}.$$

Consider, for example, the set $\mathbf{Z}$ of integers and the equivalence relation E that contains (n_1, n_2) iff $n_1 - n_2$ is an even number. (It is an equivalence relation because $n - n = 0$ is even; $n_1 - n_2$ even implies $n_2 - n_1$ even; and $n_1 - n_2$ and $n_2 - n_3$ even implies $n_1 - n_3 = (n_1 - n_2) + (n_2 - n_3)$ is even.) Then

$$[0] = \{\text{all even numbers}\}$$
$$[1] = \{\text{all odd numbers}\}.$$

Thus **Z**/E has only two elements, even though each element of the factor set is itself a set which contains infinitely many elements, and so has infinitely many labels:

$$[0] = [n] \qquad \text{iff } n \text{ is an even number}$$

$$[1] = [n] \qquad \text{iff } n \text{ is an odd number.}$$

This last pair of equations is an example of the general fact (which the reader should prove to follow from 1 and 2) that:

3 **FACT**: If E is an equivalence relation on A, then

$$[a_1] = [a_2] \qquad \text{iff } (a_1, a_2) \in E$$

$$[a_1] \cap [a_2] = \emptyset \qquad \text{iff } (a_1, a_2) \notin E.$$

□

We also note that

4 **FACT**: If E is an equivalence relation on A, then the **canonical onto map**

$$\eta_E : A \longrightarrow A/E : a \mapsto [a]_E$$

is indeed onto. □

We sometimes find it convenient to write $a_1 E a_2$ rather than $(a_1, a_2) \in E$.

We now recapture the process of going from a sorting map to equivalence classes in bijective correspondence with the barrels:

5 **DEFINITION**: Given any function $f : A \longrightarrow B$, we define the **equivalence relation** $E(f)$ **of** f on A by

$$E(f) = \{(a, a') \in A \times A \mid f(a) = f(a')\}.$$

The reader should check that $E(f)$ really is an equivalence relation. Then (recalling 4 for the case $E = E(f)$) we have

6 **PROPOSITION**: If $f : A \longrightarrow B$ maps A into B, then the map

$$k : A/E(f) \longrightarrow f(A) : [a]_{E(f)} \mapsto f(a)$$

is a bijection, and $f = i \cdot k \cdot \eta_f : A \longrightarrow B$, where $i : f(A) \longrightarrow B$ is the inclusion $b \mapsto b$, and we write η_f instead of $\eta_{E(f)}$.

Proof: Since f maps A onto $f(A)$, k must be onto. Since $E(f)$ is an equivalence relation

$$[a]_{E(f)} = [a']_{E(f)} \quad \text{iff} \quad (a, a') \in E(f) \quad \text{iff} \quad f(a) = f(a')$$

and so k is one-to-one. Thus k is a bijection. Finally,

$$k \cdot \eta_f(a) = k([a]_{E(f)}) = f(a), \qquad \text{as claimed.}$$

□

7 **COROLLARY**: f is an onto map iff $f = k \cdot \eta_f$, where k is a bijection. □

It is now time to see if we can recast equivalence relations in arrow-form. Given *any* relation R on A (i.e. any subset of $A \times A$) we may associate with it the two projection maps

$$p_k : R \longrightarrow A : (a_1, a_2) \mapsto a_k \qquad k = 1, 2.$$

On the other hand, given any pair of maps

$$p_1, p_2 : R \longrightarrow A$$

with common domain, and with codomain A, we may define on A the corresponding relation

$$E_R = \{(p_1(r), p_2(r)) \mid r \in R\} \subset A \times A \tag{1}$$

(where we forbear writing $E_{(R,\, p_1,\, p_2)}$ for brevity). This suggests the following arrow-theoretic generalization of a relation:

8 **DEFINITION**: A **relation** on the set A is a pair of maps $p_1, p_2 : R \longrightarrow A$ with common domain, and with codomain A.

The reader may wish to glimpse at exercise 1.

Given a relation $R \overset{p_1}{\underset{p_2}{\rightrightarrows}} A$, we can define the **equivalence relation generated by** (R, p_1, p_2) to be $\bar{R}$, the smallest equivalence relation (see exercise 2) containing the relation defined by R: $(a, a') \in \bar{R}$ iff either $a = a'$ or (a, a') can be linked by a chain $(a_1, a_2, \ldots, a_{n+1})$ of elements of A where $a = a_1$, $a' = a_{n+1}$ and for each k with $1 \leqslant k \leqslant n$, either (a_k, a_{k+1}) or (a_{k+1}, a_k) belongs to E_R. We leave it to the reader to prove that $\bar{R}$ is an equivalence relation.

We may then define the **canonical onto map**

$$\eta_{\bar{R}} : A \longrightarrow A/\bar{R} : a \mapsto [a]_{\bar{R}}$$

and we note that

$$\eta_{\bar{R}} \cdot p_1(r) = \eta_{\bar{R}} \cdot p_2(r) \qquad \text{for every } r \text{ in } R$$

since $(p_1(r), p_2(r)) \in \bar{R}$ for every $r \in R$. Thus we have that

$$R \overset{p_1}{\underset{p_2}{\rightrightarrows}} A \xrightarrow{\eta_{\bar{R}}} A/\bar{R} \tag{2}$$

recalling the discussion of diagram (3) in Section 1.1.

9 **LEMMA**: If $R \overset{p_1}{\underset{p_2}{\rightrightarrows}} A$ generates the equivalence relation $\bar{R}$ then the diagram

$$R \overset{p_1}{\underset{p_2}{\rightrightarrows}} A \xrightarrow{\eta_{\bar{R}}} A/\bar{R}$$

has the property that if h' satisfies

$$R \overset{p_1}{\underset{p_2}{\rightrightarrows}} A \xrightarrow{h'} B \tag{3}$$

then there is a unique $\psi : A/\bar{R} \longrightarrow B$ such that

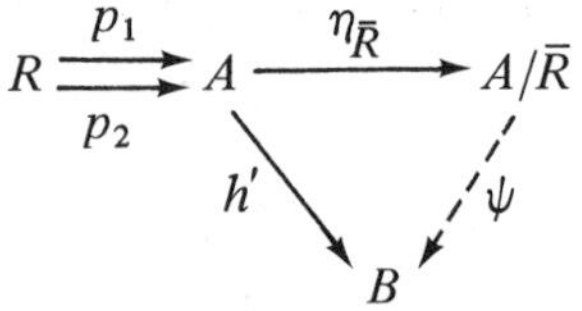

(4)

Proof: If h satisfies (3), then the map $\psi : A/\bar{R} \longrightarrow B : [a] \longrightarrow h'(a)$ is well-defined as follows: If $[a] = [a']$ then $a = a'$ or there is a chain $(a_1, \ldots, a_{n+1})$ with $a = a_1$, either (a_j, a_{j+1}) or (a_{j+1}, a_j) R-related for $1 \leqslant j \leqslant n$, and $a_{n+1} = a'$.

Now if a_j and a_k are R-related, (4) tells us that $h'(a_j) = h'(a_k)$. Thus if $[a] = [a']$, $h'(a) = h'(a_1) = \ldots h'(a_{n+1}) = h'(a')$. Moreover, ψ is the only map which makes (4) commute as $\eta_{\bar{R}}$ is onto. □

We are at last ready for a formal definition, and for two propositions that give a new arrow-characterization of onto maps.

10 DEFINITION: We say a map $A \xrightarrow{h} B$ is a **coequalizer** iff there exists a pair $p_1, p_2 : R \longrightarrow A$ of maps such that $h \cdot p_1 = h \cdot p_2$, and such that whenever $A \xrightarrow{h'} B'$ satisfies $h' \cdot p_1 = h' \cdot p_2$, there is a unique map $B \xrightarrow{\psi} B'$ such that $\psi \cdot h = h'$

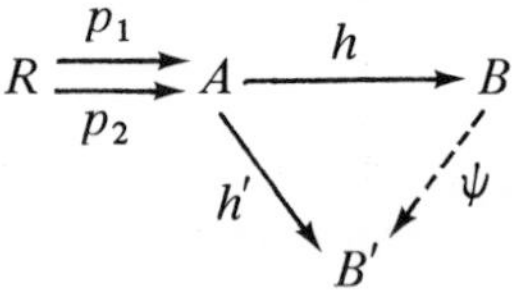

In this situation, we call h the **coequalizer of p_1 and p_2**, and write $h = \text{coeq}(p_1, p_2)$.

We immediately have

11 PROPOSITION: Every coequalizer is an epimorphism.

Proof: Suppose $h = \text{coeq}(p_1, p_2)$ and that $k_1 \cdot h = k_2 \cdot h$. We must prove that $k_1 = k_2$. But if we take $h' = k_1 \cdot h$ in 10 – which we may since $h' \cdot p_1 = k_1 \cdot (h \cdot p_1) = k_1 \cdot (h \cdot p_2) = h' \cdot p_2$ – we see that there is a *unique* ψ such that $\psi \cdot h = h'$. But $h' = k_1 \cdot h = k_2 \cdot h$, by hypothesis, and so we must have that $k_1 = \psi = k_2$. Thus h is an epimorphism. □

12 PROPOSITION: Every onto map is a coequalizer.

Proof: Simply take $R \underset{p_2}{\overset{p_1}{\rightrightarrows}} A$ to be the equivalence relation $E(f)$ of $f : A \longrightarrow B$, i.e. $R = \{(a, a') \mid a, a' \text{ in } A \text{ and } f(a) = f(a')\}$ while p_1 and p_2 are the projections. Because f is onto, we have by 7 that $f = k \cdot \eta_f$ where k is a bijection. Thus, η_f is the coequalizer of p_1 and p_2 by 9, and this ensures (why?) that $f = \operatorname{coeq}(p_1, p_2)$. □

We used the word 'epimorphism' in 11 and 'onto map' in 12 to emphasize that the proof of 11 only uses arrows, while the proof of 12 (via its dependence on 9) makes explicit use of elements.

Reversing the arrows in 10 we obtain

13 DEFINITION: We say a map $B \xrightarrow{h} A$ is an **equalizer** iff there exists a pair $q_1, q_2 : A \longrightarrow R$ of maps such that $q_1 \cdot h = q_2 \cdot h$, and such that whenever $B' \xrightarrow{h'} A$ satisfies $q_1 \cdot h' = q_2 \cdot h'$, there is a unique map $B' \xrightarrow{\phi} B$ such that $h \cdot \phi = h'$

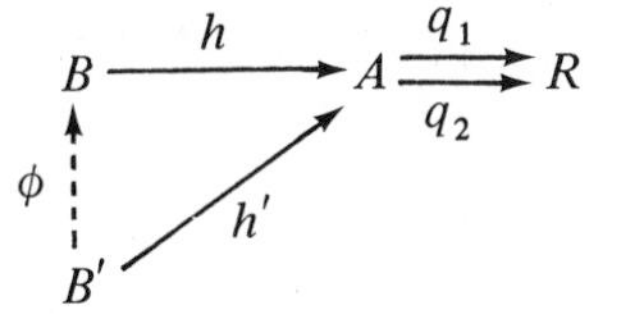

In this situation, we call h the **equalizer of** q_1 **and** q_2, and write $h = \operatorname{eq}(q_1, q_2)$.

We immediately have

14 PROPOSITION: Every equalizer is a monomorphism. □

Why is there no proof given? Re-read the discussion following the proof of 2.6, and then recall that the proof of 11 involves only arrow-chasing to see why.

Let us now find out what equalizers really look like in the category of sets: We saw that $\operatorname{coeq}(p_1, p_2)$ was (we omit the standard qualification 'up to isomorphism') the canonical onto map of the smallest equivalence relation $\bar{R}$ including the relation defined on A by $R \overset{p_1}{\underset{p_2}{\rightrightarrows}} A$. We now provide a corresponding characterization of the relationship between $\operatorname{eq}(q_1, q_2)$ and $A \overset{q_1}{\underset{q_2}{\rightrightarrows}} R$. We start by noting (14) that every equalizer is one-to-one, so that the diagram

$$B \xrightarrow{h} A \overset{q_1}{\underset{q_2}{\rightrightarrows}} R$$

suggests that we identify (remember that we relabel sets *ad libitum*!) B with the subset $h(B)$ of A with which it is in one-to-one correspondence via $b \leftrightarrow h(b)$. This subset is then characterized by the fact that $q_1(a) = q_2(a)$ for every a in it. In fact, we may show that $h(B)$ contains *every* a such that $q_1(a) = q_2(a)$. To see this, choose any a' such that $q_1(a') = q_2(a')$. To prove that $a' \in h(B)$, set $B' = B \cup \{a'\}$, and let $h' : B' \longrightarrow A$ be defined by

$$h'(b) = \begin{cases} h(b) & \text{if } b \in B \\ a' & \text{if } b = a'. \end{cases}$$

Then it is certainly true that $q_1 \cdot h' = q_2 \cdot h'$, and so there exists a unique $\phi : B' \longrightarrow B$ such that

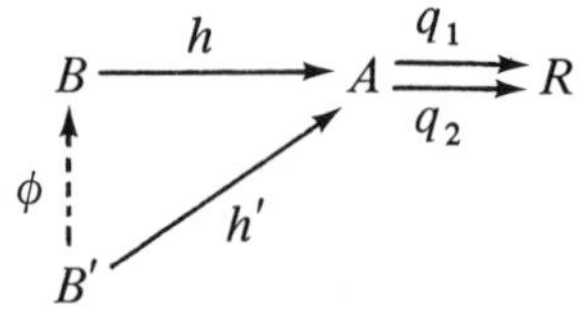

commutes. Therefore $a' \in h'(B') = h(\phi(B')) \subset h(B)$ and $a' \in h(B)$. Thus we have indeed shown that

$$h(B) = \{a \mid a \text{ in } A \text{ and } q_1(a) = q_2(a)\}.$$

In other words, $h(B)$ is the subset of A which 'equalizes' q_1 and q_2, and it is for this reason that we have the term *equalizer*, with the term *coequalizer* following by duality.

15 **LEMMA**: The equalizer $B \xrightarrow{h} A$ of $A \underset{q_2}{\overset{q_1}{\rightrightarrows}} R$ is the inclusion map

$$h : \{a \mid a \in A \text{ and } q_1(a) = q_2(a)\} \longrightarrow A, \quad a \mapsto a. \tag{5}$$

Proof: We have already seen that *if* $A \underset{q_2}{\overset{q_1}{\rightrightarrows}} R$ has an equalizer it must be given (up to isomorphism, of course) by (5). We must now check that, given *any* $A \underset{q_2}{\overset{q_1}{\rightrightarrows}} R$, (5) defines its equalizer. Suppose, then, that we define h by (5) and consider any other h' for which $q_1 \cdot h = q_2 \cdot h$:

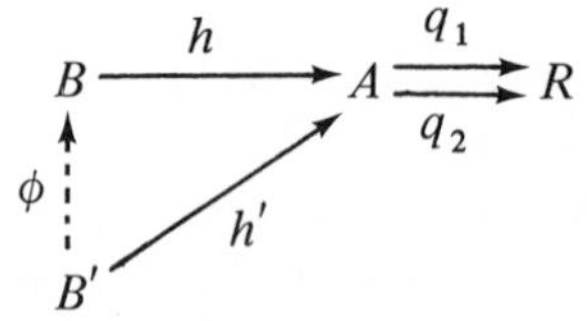

For every b' in B', $q_1(h'(b')) = q_2(h'(b'))$ and so $h'(b')$ lies in B. Thus we can take $\phi(b') = h'(b')$ for each b' in B', and it is clear that this choice of ϕ is unique, since $h(a) = a$. □

At this stage let us summarize our findings about arrow-characterizations of onto and one-to-one set maps:

16 In the category of sets:

f is a monomorphism iff f is one-to-one iff f is an equalizer

f is an epimorphism iff f is onto iff f is a coequalizer. □

This table is interesting because we shall see in later chapters that there exist categories (domains of mathematical discourse) in which there exist monomorphisms which are not equalizers, and epimorphisms which are not coequalizers.

Exercises

1 Given relations $p_1, p_2 : R \longrightarrow A$, $q_1, q_2 : S \longrightarrow A$, say that $(p_1, q_1) \sim (q_1, q_2)$ if there exist functions (not necessarily isomorphisms!) as shown below:

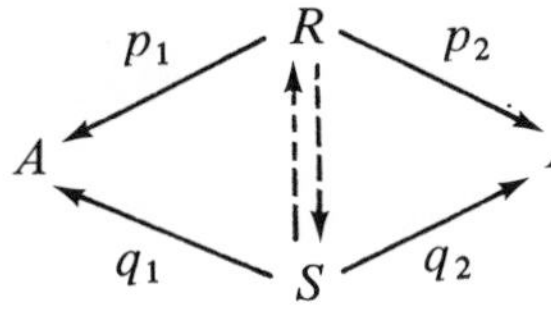

Show that $\sim$ is an equivalence relation. Show that $[R, p, q] \mapsto E_R$ is a well-defined bijection from the set of $\sim$-equivalence classes of relations to the set of subsets of $A \times A$.

2 Let $p_1, p_2 : R \longrightarrow A$ and let $\bar{R}$ be the equivalence relation generated by (R, p_1, p_2). Show that if $S \subset A \times A$ is an equivalence relation such that $E_R \subset S$ then $\bar{R} \subset S$.

3 A **partition** of a set A is a set P of non-empty subsets of A which 'divide A up into pieces', i.e. every element of A belongs to a unique element of P. Show that $E = \{(a_1, a_2) \mid a_1$ and a_2 belong to the same element of $P\}$ is an equivalence relation on A and show that $A/E \cong P$.

4 Show that every set A admits a unique equivalence relation E such that A/E has one element.

5 Let $p_1, p_2 : R \longrightarrow A$ be functions with disjoint images, i.e. $p_1(R) \cap p_2(R) = \emptyset$. Show that if B is a 1-element set and $h : A \longrightarrow B$ is the unique function then $h = \text{coeq}(p_1, p_2)$. [Hint: Use 10, not the details of 9.]

6 Given $f : A \longrightarrow B$ show that $\text{id}_A = \text{eq}(f, f)$ and that $\text{id}_B = \text{coeq}(f, f)$.

7 Given $p_1, p_2 : R \longrightarrow A$, $h : A \longrightarrow B$ prove that $h = \text{coeq}(p_1, p_2)$ iff $2^h = \text{eq}(2^{p_1}, 2^{p_2})$ (see exercise 1.4). Similarly, given $q_1, q_2 : A \longrightarrow R$, $h : B \longrightarrow A$ prove $h = \text{eq}(q_1, q_2)$ iff $2^h = \text{coeq}(2^{q_1}, 2^{q_2})$.

8 Given $f : A \longrightarrow B$, $g : B \longrightarrow A$ with $g \cdot f = \text{id}_A$, prove that $g = \text{coeq}(\text{id}_B, f \cdot g)$. Can you state and prove the dual result?

9 Give an arrow-theoretic proof that if $f : A \longrightarrow B$ is a coequalizer and a monomorphism then f is an isomorphism.

Chapter 2

BASIC CONCEPTS OF CATEGORY THEORY

We have seen that a number of basic concepts of set theory can be re-expressed as properties of maps rather than properties of elements. In particular, we have given arrow-theoretic definitions for epimorphisms, monomorphisms and isomorphisms; products and coproducts; equalizers and coequalizers. (Some readers will find it useful to write out these definitions as they occurred in Chapter 1, and then keep the list available for easy reference as we proceed through Chapter 2.) These concepts will, so formulated, be available and useful in a wide range of domains of discourse in addition to that of set theory. In this chapter we work out this implication by making the general definition of a *category* as the abstract formulation of the informal notion of *domain of mathematical discourse* and use linear algebra (with which we assume the reader has a little familiarity) and partially ordered sets (of which we give a self-contained account) to illustrate the wide applicability of our basic concepts of category theory. In the next two chapters we shall extend this survey by introducing the reader to a number of important domains of discourse from algebra, topology and analysis.

Our strategy in this chapter is as follows: In Section 1 we present the basic concepts of linear algebra and partially ordered sets that we will need. Then, in Section 2, we give the definition of a category, and verify that not only are **Set** and $\mathbf{Set}^{\mathrm{op}}$ categories, but so is the collection **Vect** of vector spaces and linear maps, and so is any partially ordered set! In the remainder, we investigate the arrow-theoretic concepts mentioned above in arbitrary categories.

2.1 VECTOR SPACES† AND POSETS

The classic example of a vector space is $\mathbf{R}^n$, which is the cartesian product of n copies of the real line $\mathbf{R}$, with the usual convention that we write a typical element as a column **vector**

† The reader all too familiar with vector spaces and linear maps should go straight to the paragraph following 4 on page 27.

$$x = \begin{bmatrix} x_1 \\ x_2 \\ \vdots \\ x_n \end{bmatrix}$$

each of whose n elements is a member of **R**. We define **addition** and **multiplication-by-a-scalar** component by component:

$$\begin{bmatrix} x_1 \\ x_2 \\ \vdots \\ x_n \end{bmatrix} + \begin{bmatrix} x_1' \\ x_2' \\ \vdots \\ x_n' \end{bmatrix} = \begin{bmatrix} x_1 + x_1' \\ x_2 + x_2' \\ \vdots \\ x_n + x_n' \end{bmatrix} \quad \text{and} \quad \lambda \cdot \begin{bmatrix} x_1 \\ x_2 \\ \vdots \\ x_n \end{bmatrix} = \begin{bmatrix} \lambda x_1 \\ \lambda x_2 \\ \vdots \\ \lambda x_n \end{bmatrix} \tag{1}$$

which we may rewrite in the convenient shorthand

$$(x + x')_j = x_j + x_j' \quad \text{and} \quad (\lambda x)_j = \lambda x_j \tag{2}$$

for all x, x' in $\mathbf{R}^n$ and all $\lambda \in \mathbf{R}$. (We call each real number a **scalar**: λx is x 'scaled-up' by a factor of λ.)

We use 0 to denote not only the number $0 \in \mathbf{R}$ but also the vector $0 \in \mathbf{R}^n$ which has every component zero. We use $-x$ to denote the vector with $(-x)_j = -x_j$ for each $j = 1, 2, \ldots, n$. We then immediately have from (1), and the properties of the real numbers, that

$$\begin{aligned} (x + x') + x'' &= x + (x' + x'') \\ x + x' &= x' + x \\ x + 0 &= x \\ x + (-x) &= 0 \end{aligned}$$

for all x, x' and x'' in $\mathbf{R}^n$. Moreover, we see from (2) that scalar multiplication has the properties

$$\begin{aligned} 1 \cdot x &= x \\ \lambda \cdot (x + x') &= \lambda \cdot x + \lambda \cdot x' \\ (\lambda + \lambda') \cdot x &= \lambda \cdot x + \lambda' \cdot x \\ \lambda \cdot (\lambda' \cdot x) &= (\lambda \cdot \lambda') \cdot x \ . \end{aligned}$$

We may say that any space which shares these properties is a vector space. (More generally, we could replace **R** by more general fields of scalars, such as the complex numbers **C**, but this adds nothing to our current discussion.)

1 DEFINITION: A (real) **vector space**, is a set X equipped with two functions called

addition: $X \times X \to X : (x, x') \mapsto x + x'$; and

multiplication-by-a-scalar: $\mathbf{R} \times X \to X : (\lambda, x) \mapsto \lambda \cdot x$

which satisfy the two following sets of properties:

I. There is given an element $0 \in X$, and, for each $x \in X$, a $-x \in X$ such that

$$(x + x') + x'' = x + (x' + x'')$$
$$x + x' = x' + x$$
$$x + 0 = x$$
$$x + (-x) = 0$$

for all x, x', x'' in X.

II.

$$1 \cdot x = x$$
$$\lambda \cdot (x + x') = \lambda \cdot x + \lambda \cdot x'$$
$$(\lambda + \lambda') \cdot x = \lambda \cdot x + \lambda' \cdot x$$
$$\lambda \cdot (\lambda' \cdot x) = (\lambda \cdot \lambda') \cdot x$$

for all λ, λ' in $\mathbf{R}$ and all x, x' in X.

We call elements of X **vectors** and elements of $\mathbf{R}$ **scalars.**

To see how to work with the abstract definition, we show that it implies $0 \cdot x = 0$ (where the first 0 is a scalar, the second a vector) for all x in X:

$$0 \cdot x + 0 \cdot x = (0 + 0) \cdot x = 0 \cdot x \ .$$

Thus

$$\begin{aligned} 0 \cdot x = 0 \cdot x + 0 &= 0 \cdot x + (0 \cdot x + [-(0 \cdot x)]) = (0 \cdot x + 0 \cdot x) + [-(0 \cdot x)] \\ &= (0 + 0) \cdot x + [-(0 \cdot x)] = 0 \cdot x + [-(0 \cdot x)] = 0. \end{aligned}$$

In the sequel, we shall feel free to use all such basic properties of vector spaces without proof, since we assume the reader has already been exposed to them. We now give two examples of vector spaces other than $\mathbf{R}^n$:

Given any set I (possibly infinite) we consider the set of functions (cf. 1.2.8)

$$\mathbf{R}^I = \{f \mid f \text{ maps } I \text{ into } \mathbf{R}\} \ .$$

For $I = \{1, 2, ..., n\}$ this is just like $\mathbf{R}^n$, where we identify each f with the column vector x with $x_j = f(j)$. With this model, we use (2) to define addition and multiplication-by-a-scalar (*componentwise* or *coordinatewise*) by

$$(f + f')(i) = f(i) + f'(i) \qquad \text{and} \qquad (\lambda \cdot f)(i) = \lambda \cdot f(i)$$

for each f and f' in $\mathbf{R}^I$, each $\lambda \in \mathbf{R}$, and each $i \in I$. It is immediate that $\mathbf{R}^I$ together with these operations is a vector space according to the conditions set forth in 1.

Given any function $f : A \longrightarrow B$ for which B contains a zero element, we call the set

$$\text{supp}(f) = \{a \mid a \in A \ \text{ and } \ f(a) \neq 0\}$$

of elements with non-zero image the **support** of f. We say that f **has finite support** if $\text{supp}(f)$ is a finite set.

Note that if f and f' are elements of $\mathbf{R}^I$, then

$$\text{supp}(f + f') \subset \text{supp}(f) \cup \text{supp}(f')$$

while $$\text{supp}(\lambda \cdot f) = \text{supp}(f) \qquad \text{unless } \lambda = 0.$$

Thus *the set of elements of* $\mathbf{R}^I$ *of finite support forms a subspace of* $\mathbf{R}^I$ where

2 The subset Y of a vector space $(X, +, \cdot)$ is called a **subspace** if

$$x + x' \in Y$$

and $$\lambda \cdot x \in Y$$

for every x, x' in Y and every λ in $\mathbf{R}$. (We say Y is **closed** under addition and multiplication-by-a-scalar.) In the obvious way, a subspace is a vector space.

We have said that a vector space is a *set with structure,* i.e. it is a set together with the two functions, addition and multiplication-by-a-scalar. Suppose we are given two vector spaces A and B. (This is shorthand for saying that we are given A equipped with two functions $A \times A \longrightarrow A : (a, a') \mapsto a \underset{A}{+} a'$ and $\mathbf{R} \times A \longrightarrow A : (\lambda, a) \mapsto \lambda \underset{A}{\cdot} a$; and that B is given with $B \times B \longrightarrow B :$ $(b, b') \mapsto b \underset{B}{+} b'$ and $\mathbf{R} \times B \longrightarrow B : (\lambda, b) \mapsto \lambda \underset{B}{\cdot} b$. However, we usually find that no confusion should arise (it may take a little practice to avoid it!) if we use + for both $\underset{A}{+}$ and $\underset{B}{+}$, and $\cdot$ for both $\underset{A}{\cdot}$ and $\underset{B}{\cdot}$, and then leave the operations implicit in speaking of vector spaces A and B rather than of vector spaces $(A, +, \cdot)$ and $(B, +, \cdot)$.) We then say a map from A to B is **linear** if it *respects the structure* on the vector spaces in the sense that it sends sums to sums and sends a scalar multiple to the multiple with the same scalar:

3 **DEFINITION**: Let A and B be vector spaces. We say a map $f : A \longrightarrow B$ is **linear** if

$$f(x + x') = f(x) + f(x') \qquad \text{and} \qquad f(\lambda \cdot x) = \lambda \cdot f(x)$$

for all x, x' in A and all λ in $\mathbf{R}$.

It will often be convenient to use the following alternate characterization:

4 The map $f : A \longrightarrow B$ of vector spaces is linear iff

$$f(\lambda \cdot x + \lambda' \cdot x') = \lambda \cdot f(x) + \lambda' \cdot f(x') \tag{3}$$

for all λ, λ' in $\mathbf{R}$ and all x, x' in A.

Proof: If f is linear, then $f(\lambda \cdot x + \lambda' \cdot x') = f(\lambda \cdot x) + f(\lambda' \cdot x') = \lambda \cdot f(x) + \lambda' \cdot f(x)$. Conversely, if f satisfies (3), then

$$f(x + x') = f(1 \cdot x + 1 \cdot x') = 1 \cdot f(x) + 1 \cdot f(x') = f(x) + f(x'), \qquad \text{while}$$
$$f(\lambda \cdot x) = f(\lambda \cdot x + 0 \cdot 0) = \lambda \cdot f(x) + 0 \cdot f(0) = \lambda \cdot f(x). \qquad \square$$

So much, for now, for vector spaces. Consider now a set S and denote the **power set** of S, i.e. the set of all subsets of S, by 2^S. The **inclusion relation** $\subset$

(where $A_1 \subset A_2$, for A_1 and A_2 subsets of S, iff $x \in A_1$ implies $x \in A_2$) has the properties

Reflexivity: $A \subset A$

Antisymmetry: $A \subset B$ and $B \subset A$ implies $A = B$

Transitivity: $A \subset B$ and $B \subset C$ implies $A \subset C$

Abstracting from this situation, we have

5 **DEFINITION:** A **poset** (short for **partially ordered set**) is a set P equipped with a relation $\leqslant$ which is

Reflexive: $p \leqslant p'$ for all p in P

Antisymmetric: $p \leqslant p'$ and $p' \leqslant p \Rightarrow p = p'$, for all p, p' in P

Transitive: $p \leqslant p'$ and $p' \leqslant p'' \Rightarrow p \leqslant p''$, for all p, p', p'' in P.

Again, a poset is a set with structure, and given two posets P and P' (where P is short for $(P, \leqslant)$, and P' is short for $(P', \leqslant)$ which in turn is short for $(P', \leqslant'))$ we may consider structure-preserving maps:

6 **DEFINITION:** Given two posets P and P', we say a map $f : P \longrightarrow P'$ is **order-preserving** if $p_1 \leqslant p_2$ implies $f(p_1) \leqslant f(p_2)$ for every p_1, p_2 in P.

For example, let S be a finite set, let 2^S be the poset of subsets of S ordered by set inclusion, and let **N** be the integers ordered under the usual 'less than or equal to' relation. Then the map

$$f : 2^S \longrightarrow \mathbf{N} : A \mapsto |A|, \qquad \text{the number of elements in } A$$

is order-preserving, since $A_1 \subset A_2$ implies $|A_1| \leqslant |A_2|$.

Exercises

1 Let $f, g : X \longrightarrow Y$ be linear. Prove that $f + g$ defined by $(f+g)(x) = f(x) + g(x)$ and $\lambda \cdot f$ defined by $(\lambda \cdot f)(x) = \lambda \cdot f(x)$ are again linear. Conclude that the linear maps between two vector spaces form a vector space.

2 A **pre-ordered set** is a set P equipped with a reflexive and transitive relation. Let $(P, \leqslant)$ be a pre-ordered set. Define $E = \{(p, q) \in P \times P : p \leqslant q$ and $q \leqslant p\}$. Show that E is an equivalence relation; that $[p] \leqslant [q]$ if $p \leqslant q$ is well-defined; that $(X/E, \leqslant)$ is a poset such that $\eta_E : (P, \leqslant) \longrightarrow (P/E, \leqslant)$ is order-preserving (definition obvious); and that

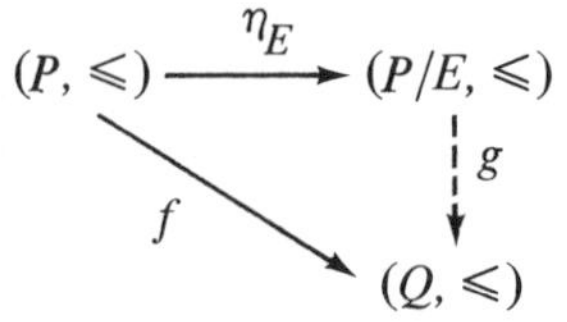

whenever $(Q, \leqslant)$ is a *poset* and $f:(P, \leqslant) \longrightarrow (Q, \leqslant)$ is order-preserving, there exists a unique order-preserving g such that $g \cdot \eta_E = f$.

3 Let X be a vector space. For $x, y \in X$ define $x \leqslant y$ to mean there exists $\lambda \geqslant 1$ with $\lambda \cdot x = y$. Show that $(X, \leqslant)$ is a poset. Show that, when vector spaces are considered as posets in this way, linear maps are order-preserving.

4 A **linearly-ordered set** is a poset $(P, \leqslant)$ with the additional property that given any $p, q \in P$, $p \leqslant q$ or $q \leqslant p$. Prove that a 3-element set admits 6 linear orders, 19 partial orders and 29 pre-orders.

5 Let $(\mathbf{R}, \leqslant')$ be partially ordered as in exercise 3; show that this is not a linear ordering: try to draw a picture of this ordering with x to the left of y whenever $x \leqslant' y$. Study the relationship between $\leqslant'$ and the usual 'less than or equal to', $\leqslant$.

6 Prove that addition and subtraction are linear maps $\mathbf{R} \times \mathbf{R} \longrightarrow \mathbf{R}$ but that $\mathbf{R} \longrightarrow \mathbf{R}$, $x \mapsto x + 1$ is not linear.

2.2 THE DEFINITION OF A CATEGORY

We now isolate some of the most elementary properties of the domain of discourse of *set theory*, and enshrine them in our general definition of a category:

1. We are given a set of *objects*, namely the sets.
2. Given any pair of objects (sets) A and B, we have the collection of all maps $f: A \longrightarrow B$ from A to B. For consistency with our general notation below, we call each $f: A \longrightarrow B$ a *morphism* from A to B, and write **Set**(A, B) for the set of all such morphisms.
3. Given $f \in$ **Set**(A, B) and $g \in$ **Set**(B, C), we may form their composite $g \cdot f \in$ **Set**(A, C).
4. For each object (set) A there is an identity morphism $\mathrm{id}_A : A \longrightarrow A$.
5. For each $f: A \longrightarrow B$, $g: B \longrightarrow C$, $h: C \longrightarrow D$ we have
$$[(h \cdot g) \cdot f](a) = [h \cdot g](f(a)) = h(g(f(a)))$$
$$= h([g \cdot f](a)) = [h \cdot (g \cdot f)](a)$$
6. For every $f: A \longrightarrow B$, we have $f \cdot \mathrm{id}_A(a) = f(a)$ and $\mathrm{id}_B \cdot f(a) = f(a)$.

These six properties of **Set** are all that we ask of any collection of 'objects' and 'morphisms' in order that it be dignified with the name 'category':

1 **DEFINITION**: A **category K** comprises a collection Obj(**K**), called the set of **objects** of **K**, together with, for each pair A, B of objects of **K**, a distinct (possibly empty) set **K**(A, B) called the set of **morphisms** from A to B subject to the conditions CAT 1 and CAT 2 below. We may write $f: A \longrightarrow B$ or $A \xrightarrow{f} B$ to indicate that the morphism f is in **K**(A, B), and we then refer to A as the **domain** of f and to B as the **codomain** of f:

CAT 1: For any three (not necessarily distinct) objects A, B and C of **K**, there are given functions

$$\mathbf{K}(A, B) \times \mathbf{K}(B, C) \longrightarrow \mathbf{K}(A, C) : (A \xrightarrow{f} B,\ B \xrightarrow{g} C) \mapsto A \xrightarrow{g \cdot f} C$$

called **composition** which satisfy the **associative axiom** that for all objects A, B, C, D of **K** and all morphisms f in $\mathbf{K}(A, B)$, g in $\mathbf{K}(B, C)$ and h in $\mathbf{K}(C, D)$ we have $h \cdot (g \cdot f) = (h \cdot g) \cdot f : A \longrightarrow D$.

CAT 2: For every object A of **K**, the set $\mathbf{K}(A, A)$ contains (possibly among other morphisms) a special morphism id_A, called the **identity** of A, with the property that for every object B of **K**, and for all $f \in \mathbf{K}(A, B)$ and $g \in \mathbf{K}(B, A)$ we have

$$A \xrightarrow{\mathrm{id}_A} A \xrightarrow{f} B = A \xrightarrow{f} B \quad \text{and} \quad B \xrightarrow{g} A \xrightarrow{\mathrm{id}_A} A = B \xrightarrow{g} A .$$

We thus write $A \xrightarrow{f} B \xrightarrow{g} C$ for $A \xrightarrow{g \cdot f} C$; the associativity axiom then says that we may write longer arrow-chains such as $A \xrightarrow{f} B \xrightarrow{g} C \xrightarrow{h} D$ without ambiguity as to the composite morphism $h \cdot g \cdot f = (h \cdot g) \cdot f = h \cdot (g \cdot f)$ which is indicated.

We say $\mathbf{K}_1$ is a **subcategory** of **K** if $\mathbf{K}_1$ is a category with $\mathrm{Obj}(\mathbf{K}_1) \subset \mathrm{Obj}(\mathbf{K})$, while $\mathbf{K}_1(A, B) \subset \mathbf{K}(A, B)$ for any objects A, B of $\mathbf{K}_1$. We say $\mathbf{K}_1$ is **full** if $\mathbf{K}_1(A, B) = \mathbf{K}(A, B)$ for any A, B in $\mathbf{K}_1$.

As an exercise in the use of these axioms, note that only one morphism in each $\mathbf{K}(A, A)$ can be an identity, for if $k : A \longrightarrow A$ satisfies even the weak version of the identity axiom that $k \cdot f = f$ for all $f \in \mathbf{K}(A, A)$, we have that $k \cdot \mathrm{id}_A = \mathrm{id}_A$, while the axiom for id_A says that $k \cdot \mathrm{id}_A = k$, so that $k = \mathrm{id}_A$.

We have already seen that ⟨Sets and Ordinary Set Maps⟩ constitute a category **Set**. In the remainder of this section we not only verify that ⟨Vector Spaces and Linear Maps⟩ constitute a category **Vect** and that ⟨Posets and Order-Preserving Maps⟩ constitute a category **Poset**, but shall also see that each individual poset P may itself be considered as a category! We shall then close the section by defining the opposite $\mathbf{K}^{\mathrm{op}}$ of any category **K**, generalizing the construction of $\mathbf{Set}^{\mathrm{op}}$ of Section 1.1. In the remainder of this chapter we shall present the general categorical definitions of monomorphisms, epimorphisms, isomorphisms, products, etc., and study their special form in **Vect**, **Poset** and poset P *qua* category. Then in the next two chapters we shall introduce the reader to a variety of other categories such as groups and metric spaces giving a self-contained exposition of each new concept as we come to it.

2 *The Category* Vect

We have already specified that **Vect** shall have *vector spaces* for objects, and that for each pair of objects A and B, $\mathbf{Vect}(A, B)$ shall be the set of *linear* maps from A into B.† Let us define composition and identities as in **Set** by $A \xrightarrow{f} B \xrightarrow{g} C : a \mapsto (g \cdot f)(a) = g(f(a))$ and $\mathrm{id}_A : A \longrightarrow A : a \mapsto a$. We must then verify that if g and f are linear, then so too is $g \cdot f$, and that id_A is always linear:

If f and g are linear

$$\begin{aligned}(g \cdot f)(\lambda \cdot a + \lambda' \cdot a') &= g[f(\lambda \cdot a + \lambda' \cdot a')] \\ &= g[\lambda \cdot f(a) + \lambda' \cdot f(a')] && \text{since } f \text{ is linear} \\ &= \lambda \cdot g(f(a)) + \lambda' \cdot g(f(a')) && \text{since } g \text{ is linear} \\ &= \lambda \cdot (g \cdot f)(a) + \lambda' \cdot (g \cdot f)(a')\end{aligned}$$

so that $g \cdot f$ is also linear. Of course, the linearity of each id_A is trivial! But since composition and identies are defined as in **Set** it is clear that they must satisfy the associativity and identity axioms of CAT 1 and CAT 2 of 1. Thus **Vect** is indeed a category.

3 *The Category* **Poset**

We have already specified that **Poset** shall have *posets* for objects, and that for each pair of objects A and B, **Poset**(A, B) shall be the set of *order-preserving* maps from A into B. Let us define composition and identities as in **Set**. We must then verify that if g and f are order-preserving, then so too is $A \xrightarrow{f} B \xrightarrow{g} C$ and that id_A is always order-preserving.

If f and g are order-preserving, then

$$\begin{aligned} a \leqslant a' \text{ in } A &\Rightarrow f(a) \leqslant f(a') \text{ in } B && (f \text{ is order-preserving}) \\ &\Rightarrow g(f(a)) \leqslant g(f(a')) \text{ in } C && (g \text{ is order-preserving})\end{aligned}$$

so that $g \cdot f$ is also order-preserving. id_A is clearly order-preserving. But since composition and identities are defined as in **Set**, we see that **Poset** is indeed a category.

The comparison of 2 and 3 cries out for a unification:

4 *NAIVE APPROACH TO CATEGORIES OF SETS WITH STRUCTURE.*

Let us be given a type of **structure** which may be placed on sets (such as '$a = \lambda_1 \cdot a_1 + \lambda_2 \cdot a_2$' in the case of vector spaces; and '$a_1 \leqslant a_2$' in the case of posets), and let us say a map $f : A \longrightarrow B$ from a set A with this type of structure to a set B with the same type of structure is **structure-preserving** if, when we replace each element a in a structure in A by $f(a)$ we get the corresponding structure in B (so that '$a = \lambda_1 \cdot a_1 + \lambda_2 \cdot a_2$' yields '$f(a) = \lambda_1 \cdot f(a_1) + \lambda_2 \cdot f(a_2)$' for

† This is the category in which linear algebra lives. We could also form a category **S-Vect**, say, whose objects are still vector spaces, but such that **S-Vect**(A, B) contains *all* maps from A to B, and not just linear maps. But that is another category.

vector spaces and '$a_1 \leqslant a_2$' yields '$f(a_1) \leqslant f(a_2)$' for posets). Then ⟨Sets with the Given Structure and Structure Preserving Maps⟩ form a category with composition and identities defined as in **Set**.

However, we shall see in Section 6.1 that 4 puts the accent on the wrong syl·la′ble, and that it is easy to directly axiomatize admissible maps.

By contrast, our next example makes it clear that the objects of a category need not be sets with structure:

5 *A Poset is a Category*

Whereas in 3 we studied the category **Poset** whose objects were posets, we now fix a *single* poset $(P, \leqslant)$ and construct a category **K** whose objects are the *elements* of P. We then define the morphism set $\mathbf{K}(p, q)$ for each pair of objects (i.e. p, q are elements of P) by

$$\mathbf{K}(p, q) = \begin{cases} \{pq\}, & \text{a set with a single element which we call } pq \text{ if } p \leqslant q. \\ \emptyset, & \text{the empty set if it is false that } p \leqslant q. \end{cases}$$

(We do *not* think of $pq : p \longrightarrow q$ as a function – after all, p isn't a set. Here we may think of a morphism as a proposition: pq is the proposition that $p \leqslant q$. The point of this example is that category theory is more general than the study of 'structure-preserving' functions.)

We must next define composition

$$\mathbf{K}(p, q) \times \mathbf{K}(q, r) \longrightarrow \mathbf{K}(p, r).$$

If either of $\mathbf{K}(p, q)$ or $\mathbf{K}(q, r)$ is empty, there is nothing to do (why?). If they are both nonempty, we have $p \leqslant q$ and $q \leqslant r$ which implies, by transitivity, that $p \leqslant r$ so that we may make the (only possible) definition

$$qr \cdot pq = pr.$$

Again, since reflexivity tells us that $p \leqslant p$ so that $\mathbf{K}(p, p)$ is nonempty for each $p \in P$, we may make the (only possible) definition

$$\mathrm{id}_p = pp.$$

It is then easy to see that the axioms are satisfied:

$$qq \cdot pq = pq \qquad \text{and} \qquad pq \cdot pp = pq$$

while

$$rs \cdot (qr \cdot pq) = ps = (rs \cdot qr) \cdot pq.$$

Thus $(P, \leqslant)$ is indeed a category with set of objects P, and with morphisms the statement of the relation $\leqslant$.

To close the section we generalize the discussion (which the reader may wish to review before proceeding further) of the construction of $\mathbf{Set}^{\mathrm{op}}$.

6 Given a category **K** we define its **opposite** (or **dual**) **category** $\mathbf{K}^{\mathrm{op}}$ by

$$\text{Obj}(\mathbf{K}^{\text{op}}) = \text{Obj}(\mathbf{K})$$

while
$$\mathbf{K}^{\text{op}}(A, B) = \{A \xleftarrow{f} B \mid f \in \mathbf{K}(B, A)\}$$

with composition defined by $A \xleftarrow{f} B \xleftarrow{g} C = C \xrightarrow{f \cdot g} A$ and identities $A \xleftarrow{\text{id}_A} A$ as in $\mathbf{K}$. Since the axioms of CAT 1 and CAT 2 of 1 are preserved under arrow-reversal it is immediate that $\mathbf{K}^{\text{op}}$ is indeed a category. Note that $(\mathbf{K}^{\text{op}})^{\text{op}} = \mathbf{K}$.

For example, if $\mathbf{K}$ is the category of the poset $(P, \leqslant)$, then $\mathbf{K}^{\text{op}}$ is the category of the poset $(P, \leqslant)$ with reversed ordering: $p \geqslant q$ iff $q \leqslant p$.

We then state the duality principle which guided so much of our work in Chapter 1.

7 *DUALITY PRINCIPLE FOR CATEGORY THEORY*

Let W be any construct defined for any category $\mathbf{K}$. Then the dual of W, called co-W, is the construct defined for any category $\mathbf{K}$ by defining W in $\mathbf{K}^{\text{op}}$ and reversing all the arrows.

If T is a theorem true for all categories $\mathbf{K}$, then the dual of T, obtained by reversing all the arrows of T, is true for all categories $\mathbf{K}^{\text{op}}$, and so (since $(\mathbf{K}^{\text{op}})^{\text{op}} = \mathbf{K}$) is true for all categories.

In other words, duality 'cuts the work in half'.

Exercises

1 A category $\mathbf{K}$ is **discrete** if every morphism in $\mathbf{K}$ is an identity map. Show that if $\mathbf{K}$ is discrete then $\mathbf{K} = \mathbf{K}^{\text{op}}$. Observe that every discrete category is a poset.

2 Let X be a vector space and let $\leqslant$ be the partial ordering on X of exercise 1.3. Verify that the following data define a category $\mathbf{K}$. The objects of $\mathbf{K}$ are the elements of X. For x, y in X, $\mathbf{K}(x, y) = \{\lambda : \lambda x \leqslant y\}$. Composition is ordinary multiplication. Observe that there are many pairs of objects (x, y) such that $\mathbf{K}(x, y) = \emptyset$.

3 Let $\mathbf{K}$ be any category and let $\mathbf{C}$ be any subset of $\text{Obj}(\mathbf{K})$. Show that $\text{Obj}(\tilde{\mathbf{C}}) = \mathbf{C}$, $\tilde{\mathbf{C}}(A, B) = \mathbf{K}(A, B)$ (for $A, B \in \mathbf{C}$), with composition and identities as in $\mathbf{K}$, defines a category $\tilde{\mathbf{C}}$, the **full subcategory induced by** $\mathbf{C}$. In normal usage we drop the tilde and say "$\mathbf{C}$ is a full subcategory".

2.3 EPIMORPHISMS AND MONOMORPHISMS

We now set down some familiar definitions in an arbitrary category. Particularizations to special categories are by-and-large worked out in the exercises.

1 **DEFINITION**: A morphism f in the category **K** is an **epimorphism** if for every pair g, h of morphisms in **K** with $g \cdot f = h \cdot f$ we must have $g = h$.

This is somewhat long-winded, and so we omit mention of the category **K** when it seems safe to do so. Thus the dual of 1 is simply

2 **DEFINITION**: A morphism f is a **monomorphism** if $f \cdot g = f \cdot h$ always implies $g = h$.

Just as in our discussion of functions in Section 1.1, as soon as we write $f \cdot g = f \cdot h$ we have implied codomain h = domain f = codomain g, and domain g = domain $f \cdot g$ = domain h, so no explicit mention of these objects is required in 2.

3 **DEFINITION**: A morphism $A \xrightarrow{h} B$ is a **coequalizer** iff there exists a pair $p_1, p_2 : R \longrightarrow A$ of morphisms such that $h \cdot p_1 = h \cdot p_2$, and such that whenever $A \xrightarrow{h'} B'$ satisfies $h' \cdot p_1 = h' \cdot p_2$, there is a unique morphism† $B \xrightarrow{\psi} B'$ such that $\psi \cdot h = h'$.

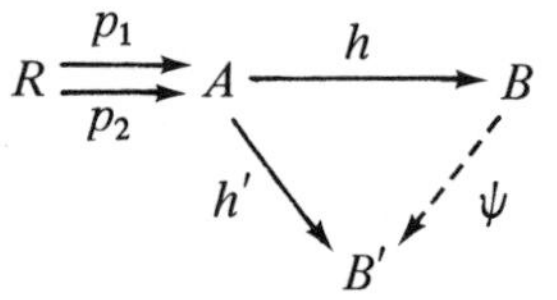

In this situation, we call h the **coequalizer of** p_1 **and** p_2 and write $h = \mathrm{coeq}(p_1, p_2)$.

As an exercise in duality, we write

4 **DEFINITION**: A morphism $B \xrightarrow{h} A$ is an **equalizer** in **K** if it is a coequalizer in $\mathbf{K}^{\mathrm{op}}$. If $h = \mathrm{coeq}(q_1, q_2)$ in $\mathbf{K}^{\mathrm{op}}$, we write $h = \mathrm{eq}(q_1, q_2)$ in **K**, and say h is the **equalizer of** q_1 **and** q_2.

Of course, by reversing the arrows, we see that $h = \mathrm{eq}(q_1, q_2)$ unpacks in **K** to the form familiar from 1.3.13: $q_1 \cdot h = q_2 \cdot h$, and whenever $B' \xrightarrow{h'} A$ satisfies $q_1 \cdot h' = q_2 \cdot h'$, there is a unique morphism $B' \xrightarrow{\phi} B$ such that $h \cdot \phi = h'$

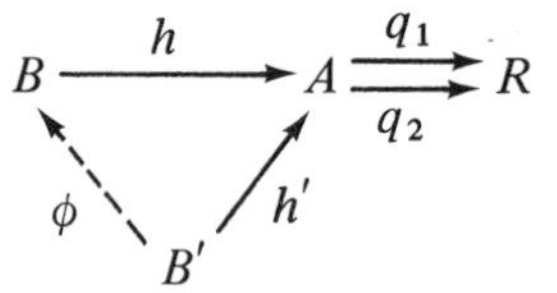

---

† Unique in the category **K** with which we are working, of course.

Just as we obtained the propositions 1.3.11 and 1.3.14 so do we have:

5 **PROPOSITION**: Every coequalizer is an epimorphism.

Proof: The proof of 1.3.11 goes through unchanged. But since it is time to pick up the tempo a little, we restate it more succinctly:

If $h = \text{coeq}(p_1, p_2)$ and $k_1 \cdot h = k_2 \cdot h$, then

$$(k_1 \cdot h) \cdot p_1 = k_1 \cdot (h \cdot p_1) = k_1 \cdot (h \cdot p_2) = (k_1 \cdot h) \cdot p_2 \qquad (1)$$

and so there is a unique morphism ψ such that $k_1 \cdot h = \psi \cdot h$. Thus $k_1 = \psi = k_2$. □

Note that the above proof makes explicit use of the associativity axiom. Since this axiom is so fundamental, we often use it without explicit mention, and simply write $k_1 \cdot h \cdot p_1 = k_1 \cdot h \cdot p_2$ for (1).

6 **PROPOSITION**: Every equalizer is a monomorphism. □

7 **DEFINITION**: A morphism $f : A \longrightarrow B$ is an **isomorphism** iff there exists a morphism $k : B \longrightarrow A$ such that $k \cdot f = \text{id}_A$ and $f \cdot k = \text{id}_B$. We call such a k an **inverse** of f. We say A and B are **isomorphic**, $A \cong B$, just in case $\mathbf{K}(A, B)$ contains an isomorphism.

8 **FACT**: An isomorphism $f : A \longrightarrow B$ is both a monomorphism and an epimorphism.

Proof: If $f : A \longrightarrow B$ is an isomorphism with inverse k, then f is a monomorphism since if $f \cdot g = f \cdot h$ then $g = \text{id}_B \cdot g = k \cdot f \cdot g = k \cdot f \cdot h = \text{id}_B \cdot h = h$. f is also an epimorphism by duality (since a coisomorphism is an isomorphism). □

As we shall see below in 12, the converse of 8 does *not* hold in every category.

9 **DEFINITION**: For fixed B, consider the class $\mathbf{M}(B)$ of all (A, f) with $f : A \longrightarrow B$ a monomorphism. Say that $(A, f) \sim (A', f')$ if

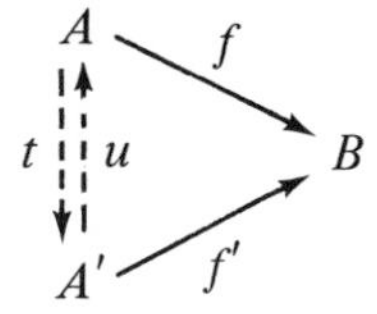

there exist $t : A \longrightarrow A'$ with $f' \cdot t = f$ and u with $f \cdot u = f'$. It is obvious that $\sim$ is an equivalence relation on $\mathbf{M}(B)$. An equivalence class $[A, f]$ is called a **mono-subobject** of B.

In **Set**, mono-subobjects of B are in bijective correspondence with subsets of B. To see this, let $f : A \longrightarrow B$ be a monomorphism and let $e : A \longrightarrow S$,

$i: S \longrightarrow B$ be the usual epi-mono factorization (1.1.15) of f where $S = f(A) \subset B$ and i is the inclusion map. Since e is onto

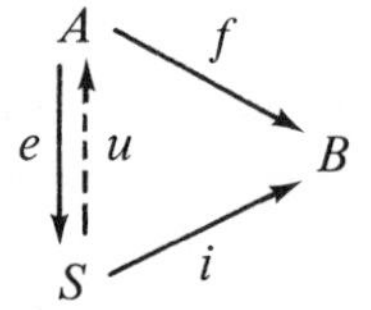

there exists for each $s \in S$ a $u(s) \in A$ such that $e(u(s)) = s$, i.e. $e \cdot u = \mathrm{id}_S$. But then $i = i \cdot e \cdot u = f \cdot u$. We leave the remaining details that the passage $[A, f] \mapsto S$ is a bijection from $\mathbf{M}(B)/\sim$ to 2^S as an exercise.

Dually, let $\mathbf{E}(B)$ be the class of all (A, f) with $f: B \longrightarrow A$ an epimorphism. Say that $(A, f) \sim (A', f')$ if there exist t, u

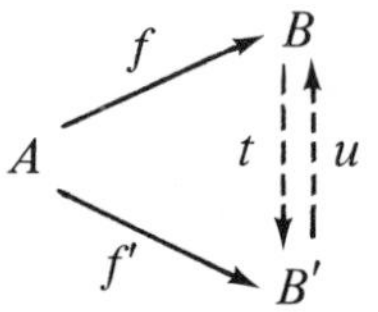

A $\sim$-equivalence class is called an **epi-quotient object** of B.

10 DEFINITION: $f: A \longrightarrow B$ is called a **retraction** or a **split epimorphism** if there exists a morphism $g: B \longrightarrow A$ such that $f \cdot g = 1_B$; and we then say that g is a **coretraction** or a **split monomorphism**, and that B is a **retract** of A.

Note that f is an isomorphism iff it is both a retraction and a coretraction. It is clear that if f is a retraction it is an epimorphism ($h_1 \cdot f = h_2 \cdot f$ implies $h_1 = h_1 \cdot 1_B = h_1 \cdot f \cdot g = h_2 \cdot f \cdot g = h_2$). Coretractions (which are monomorphisms) are also called **sections**, as motivated by the figure for the case of, e.g., **Set**:

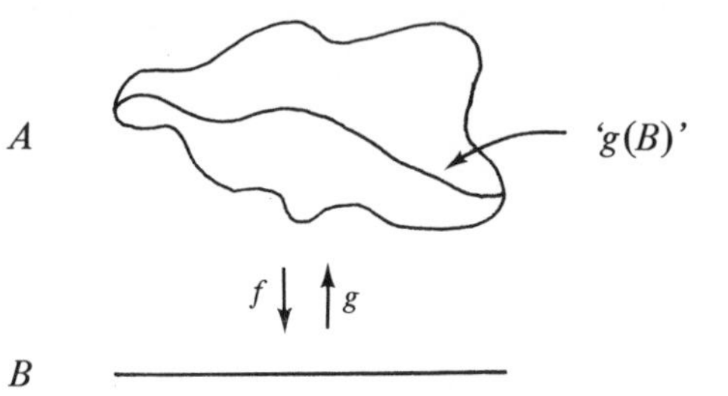

Paralleling 1.3.16 for **Set** we have

11 PROPOSITION: In the category **Vect**

(i) f is an epimorphism iff f is onto iff f is a coequalizer.

(ii) f is a monomorphism iff f is one-to-one iff f is an equalizer.

Proof: We prove (i), leaving (ii) (which is easier) to the reader.

A coequalizer is an epimorphism. By 5.

If $f : A \longrightarrow B$ is an epimorphism, f is onto. Let E be the equivalence relation $\{(b, b') \mid b - b' \in f(A)\}$ on B. Following a standard if sloppy convention, we write the quotient set as $B/f(A)$ rather than B/E. B/E is easily checked to be a vector space with operations $[b] + [b'] = [b + b']$, $\lambda \cdot [b] = [\lambda \cdot b]$. Define the two linear maps

$$t : B \longrightarrow B/f(A) : b \mapsto [b]$$
$$u : B \longrightarrow B/f(A) : b \mapsto [0] .$$

Then $t \cdot f$ and $u \cdot f$ are both zero maps (since $[f(a)] = [0]$ for all a in A) and so $t = u$ because f is an epimorphism. Thus for all $b \in B$, $[b] = [0]$, $b \in f(A)$, and f is onto.

Every onto map is a coequalizer. Given $f : A \longrightarrow B$ which is onto, form the equivalence relation

$$E = E(f) = \{(x, y) \mid x, y \text{ in } A \text{ with } f(x) = f(y)\}$$

and let p, q be the projections $(x, y) \mapsto x$ and $(x, y) \mapsto y$. Then E is easily seen to be a subspace of the well-known product vector space $A \times A$ (cf. 4.3) and p, q are linear.

Then given the diagram

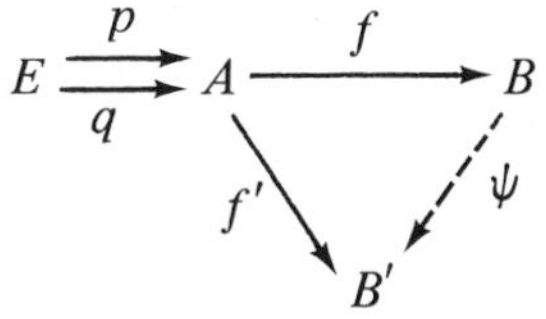

we see that $\psi : B \longrightarrow B' : f(x) \mapsto f'(x)$ is well-defined since whenever $f' \cdot p = f' \cdot q$, $f(x) = f(y)$ implies $f'(x) = f'(y)$. ψ is clearly linear:

$$\lambda \cdot f(a) + \lambda' \cdot f(a') = f(\lambda \cdot a + \lambda' \cdot a') \mapsto f'(\lambda \cdot a + \lambda' \cdot a') = \lambda \cdot f'(a) + \lambda' \cdot f'(a). \ \square$$

12 Consider now the category **K** of the poset $(P, \leqslant)$:

Every morphism is both an epimorphism and a monomorphism since each nonempty $\mathbf{K}(p_1, p_2)$ contains exactly one element; though $f : p_1 \longrightarrow p_2$ is an isomorphism iff $p_1 = p_2$, since were f to have an inverse $k : p_2 \longrightarrow p_1$ we would have $p_1 \leqslant p_2$ and $p_2 \leqslant p_1$, so that $p_1 = p_2$ by antisymmetry. Thus, for any poset P with at least one pair of distinct elements $p_1 \leqslant p_2$, we have an example of a category in which the converse of 8 fails; i.e., *a morphism may be both epi and mono without being an isomorphism* [henceforth we abbreviate "f is a monomorphism" to "f is mono", etc.].

Now let us ponder the coequalizer diagram, appropriately relabelled for $(P, \leqslant)$:

In 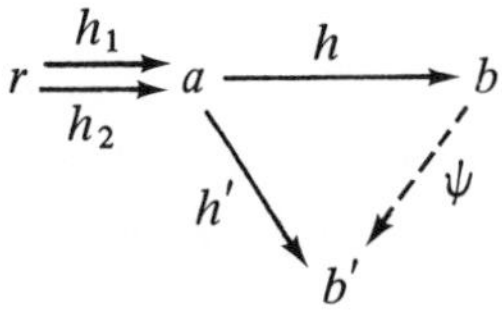 we must have $h_1 = h_2$ so that the condition

$h' \cdot h_1 = h' \cdot h_2$ is vacuous. The condition simply says that $a \xrightarrow{h} b$ (i.e. $a \leqslant b$) is a coequalizer iff $a \leqslant b'$ implies $b \leqslant b'$, for all $b' \geqslant a$. But taking $b' = a$, we deduce that $b \leqslant a$ and so (antisymmetry again) $a = b$. In other words, coequalizers are identity maps. Dually, equalizers are identity maps. Thus any poset $(P, \leqslant)$ with at least one pair of distinct elements $p_1 \leqslant p_2$ yields *a category in which there exist monomorphisms which are not equalizers, and epis which are not coequalizers* even though 5 and 6 are true in every category.

It has been gradually discovered by category theorists that no one concept of 'epimorphism' or 'monomorphism' is adequate – rather, it is profitable to axiomatize a class of possibilities.

13 DEFINITION: An **image factorization system** for a category **K** consists of a pair **(E, M)** where **E** and **M** are classes of morphisms in **K** satisfying the following four axioms:

IFS 1: If $e : A \longrightarrow B \in \mathbf{E}$ and $e' : B \longrightarrow C \in \mathbf{E}$ then $e' \cdot e : A \longrightarrow C \in \mathbf{E}$. Dually, if $m : A \longrightarrow B \in \mathbf{M}$ and $m' : B \longrightarrow C \in \mathbf{M}$ then $m' \cdot m : A \longrightarrow C \in \mathbf{M}$.

IFS 2: If $e : A \longrightarrow B \in \mathbf{E}$, e is an epimorphism. Dually, if $m : A \longrightarrow B \in \mathbf{M}$, m is a monomorphism.

IFS 3: If $f : A \longrightarrow B$ is an isomorphism then $f \in \mathbf{E}$ and $f \in \mathbf{M}$.

IFS 4: Every $f : A \longrightarrow B$ in **K** has an **E-M** factorization which is unique up to isomorphism. More precisely, there exists an **E-M factorization** (e, m) of f, meaning $e \in \mathbf{E}$, $m \in \mathbf{M}$ and $f = m \cdot e$, (so that there exists an object – call it $f(A)$ – such that e has the form $e : A \longrightarrow f(A)$ and m has the form $m : f(A) \longrightarrow B$), and this factorization is unique in the sense that if (e', m') is another such factorization – $f = m' \cdot e'$, $e' \in \mathbf{E}$,

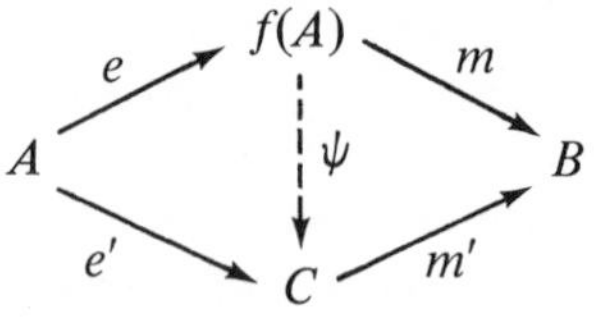

$m' \in \mathbf{M}$ – then there exists an isomorphism ψ (as shown above) with $\psi \cdot e = e'$, $m' \cdot \psi = m$.

In the category **Set** of sets,

$$\mathbf{E} = \{\text{onto functions}\} \text{ and } \mathbf{M} = \{\text{one-to-one functions}\}$$

is an image factorization system. The first three axioms are clear. For IFS 4, use 1.1.16.

14 DIAGONAL FILL-IN LEMMA: Let (**E**, **M**) be an image factorization system. Given a commutative square $- g \cdot e = m \cdot f -$ as shown below with $e \in \mathbf{E}$

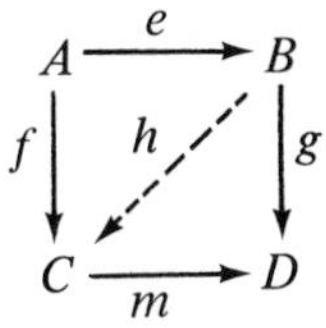

and $m \in \mathbf{M}$ there exists a unique h (as shown) with $h \cdot e = f$, $m \cdot h = g$.

Proof: Consider the diagram:

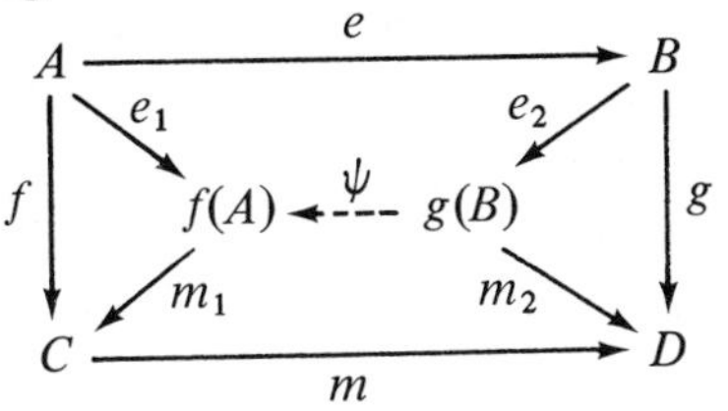

Define $h = m_1 \cdot \psi \cdot e_2$ where ψ exists using IFS 1, IFS 4. The uniqueness of h follows easily from IFS 2. □

Exercises

1 Prove that a retraction is a coequalizer. [Hint: If $f \cdot g = \mathrm{id}_B$, $f = \mathrm{coeq}(\mathrm{id}_A, g \cdot f)$]. What is the dual theorem?

2 Let E be the equivalence relation on **N** whose two equivalence classes are "even" and "odd". Note that E is a poset via $(n, m) \leqslant (n', m')$ iff $n \leqslant n'$ and $m \leqslant m'$ such that the restricted projections $p, q : E \longrightarrow \mathbf{N}$ are order-preserving. Show that $\mathrm{coeq}(p, q)$ exists in **Poset**, and is not the same as in **Set**.

3 Let X be a vector space. If A is a subset of X, show that there is at most one vector space structure on A such that the inclusion map $i : A \longrightarrow X$ is linear, and that this occurs iff A is a subspace of X. If E is an equivalence relation on X, show that there is at most one vector space structure on X/E such that $\eta_E : X \longrightarrow X/E$ is linear and that this occurs iff $[0]$ is a subspace of X (in which case, $E = \{(x, x') \mid x - x' \in [0]\}$).

4 Let $f, g : X \longrightarrow Y$ be linear. Using exercise 3, show that $\mathrm{eq}(f, g)$ and

coeq(f, g) exist in **Vect**.

5 A **preordered set** is a pair $(P, \leqslant)$ where $\leqslant$ is a reflexive and transitive (not necessarily antisymmetric) relation on P. Show that $(P, \leqslant)$ may be considered as a category as in 2.5; show that $(P, \leqslant)$ is a poset iff every isomorphism is an identity map.

6 Using 11, prove that for B in **Vect**, mono-subobjects of B may be identified with subspaces of B. [Hint: Start by completing the proof (following 9) of the corresponding result in **Set**.]

7 Prove that **E** = all linear onto maps, **M** = all linear one-to-one maps forms an image factorization system for **Vect**.

8 Prove that every linear onto map is a retraction in **Vect**. [Hint: Every vector space has a basis.]

9 Let **K** be a category in which every morphism factors as a coequalizer followed by a monomorphism. Prove that **E** = coequalizers, **M** = monomorphisms yields an image factorization system in **K**. What is the dual result?

10 An order-preserving map $f : (X, \leqslant) \longrightarrow (X', \leqslant)$ is **optimal** if for all $x_1, x_2 \in X$, $x_1 \leqslant x_2$ iff $f(x_1) \leqslant f(x_2)$. Prove that (onto, optimal one-to-one) is an image factorization system for **Poset**. Prove that, in **Poset**, 'monomorphism' coincides with 'one-to-one' whereas 'equalizer' coincides with 'one-to-one optimal'. Also prove that 'epimorphism' coincides with 'onto'. Thus the above image factorization system is (epimorphisms, equalizers).

11 Let **K** be the category of *infinite* sets and functions. Construct examples to show that a pair of morphisms in **K** need not have an equalizer or a coequalizer. [Caution: *a priori*, such constructions could exist in **K** without coinciding with the same construction relative to **Set**; e.g. if $f, g : A \longrightarrow B$ are such that $\{a \in A \mid f(a) = g(a)\}$ is finite, the argument that eq(f, g) does not exist in **K** is not yet complete.]

12 Show that in an image factorization system $(\mathbf{E}, \mathbf{M})$, **E** and **M** uniquely determine each other by showing specifically that, for fixed **E**, **M** is precisely the set of all $m : C \longrightarrow D$ such that whenever m is embedded in a

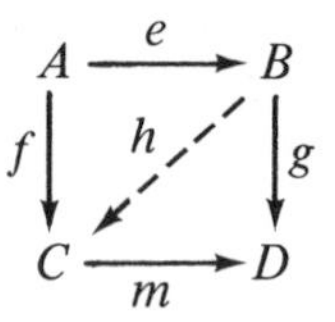

commutative square as shown above, with $e \in \mathbf{E}$, then there exists h with $h \cdot e = f$, $m \cdot h = g$.

13 (*3 × 3 lemma*). Consider the diagram

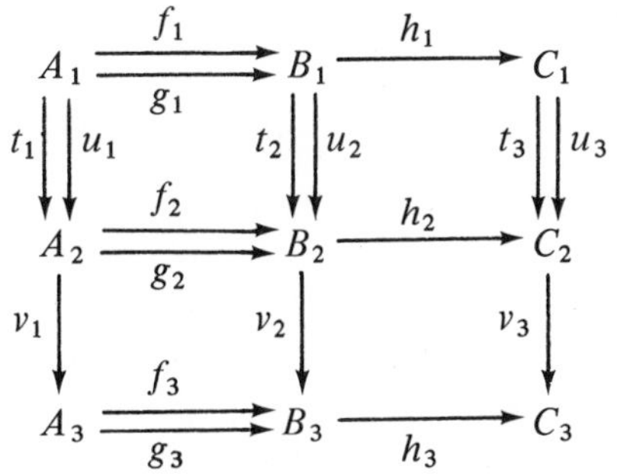

in which the three columns and top two rows are coequalizers and in which the following additional commutativities hold: $t_2 \cdot f_1 = f_2 \cdot t_1$, $u_2 \cdot g_1 = g_2 \cdot u_1$, $h_2 \cdot t_2 = t_3 \cdot h_1$, $h_2 \cdot u_2 = u_3 \cdot h_1$, $f_3 \cdot v_1 = v_2 \cdot f_2$, $g_3 \cdot v_1 = v_2 \cdot g_2$, $h_3 \cdot v_2 = v_3 \cdot h_2$. Prove that the bottom row is a coequalizer.

2.4 LIMITS AND COLIMITS

We now give the definitions of product and coproduct, generalizing 1.2.9 and exercise 1.2.3:

1 **DEFINITION**: A **product** in the category **K** of a family of objects $(A_i \mid i \in I)$ is an object A together with a family

$$(A \xrightarrow{\pi_i} A_i \mid i \in I)$$

of morphisms (called **projections**) with the property that given any other object C similarly equipped with an I-indexed family $(f_i : C \longrightarrow A_i)$ there exists a

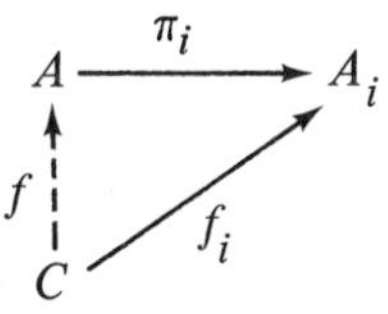

(1)

unique morphism $f : C \longrightarrow A$ such that $\pi_i \cdot f = f_i$ for all i in I. We write $A = \prod_{i \in I} A_i$. If $I = \{1, ..., n\}$ we may write $A = A_1 \times ... \times A_n$.

Dually, a **coproduct** of $(A_i \mid i \in I)$ is a family

$$(A_i \xrightarrow{\text{in}_i} A \mid i \in I)$$

(of **injections**) which, considered in $\mathbf{K}^{\text{op}}$

$$(A \xleftarrow{\text{in}_i} A_i \mid i \in I)$$

is a product there; that is, given $f_i : A_i \longrightarrow C$ there exists a unique $f : A \longrightarrow C$ such that

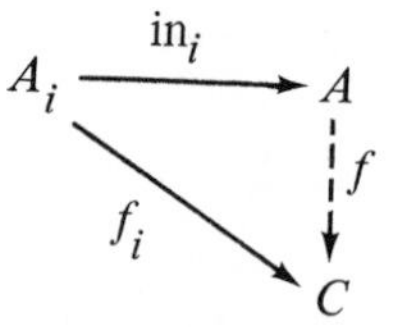

(2)

$f \cdot \mathrm{in}_i = f_i$ for all i. We write $A = \coprod_{i \in I} A_i$ and, if $I = \{1, ..., n\}$, $A = A_1 + ... + A_n$.

2 **PROPOSITION**: Let $(A_i \mid i \in I)$ be a family of sets. Then $\pi_j : \Pi A_i \longrightarrow A_j$ as defined in 1.2.8 is a product of (A_i) in **Set**. The **disjoint union**

$$\amalg A_i = \{(a, i) \mid a \in A_i\}$$

with injections

$$A_j \xrightarrow{\mathrm{in}_j} \amalg A_i : a \mapsto (a, j)$$

is a coproduct of (A_i) in **Set**.

Proof: The detailed arguments for $A_1 \times A_2$ and $A_1 + A_2$ provided in Section 1.2 generalize without substantial change. □

Despite the name, the so-called injections in_j of 1 need not be monomorphisms, although this is not common (see exercise 12).

The product of vector spaces is a cartesian product, just as in **Set**, but the coproduct is strikingly different from the disjoint union:

3 **PROPOSITION**: Given a family $(A_i \mid i \in I)$ of vector spaces, we define their **product** to be the cartesian product

$$\prod_{i \in I} A_i = \{f \mid f : I \longrightarrow \bigcup_{i \in I} A_i \text{ and } f(i) \in A_i \text{ for each } i\}$$

(as in 1.2.8) equipped with the 'coordinatewise' addition and multiplication-by-a-scalar

$$(f + f')(i) = f(i) + f'(i) \quad \text{and} \quad (\lambda \cdot f)(i) = \lambda \cdot f(i)$$

for all f, f' in $\prod_{i \in I} A_i$, λ in **R** and $i \in I$. Then $\prod_{i \in I} A_i$ together with the projections

$$\pi_j : \prod_{i \in I} A_i \longrightarrow A_j : f \mapsto f(j)$$

is a product of (A_i) in the category **Vect** in the sense of 1.

Proof: Just check that $p(c)(i) = p_i(c)$ does the trick in (1). Don't forget to check that π_j and p are linear! □

4 **PROPOSITION**: Given a family $(A_i \mid i \in I)$ of vector spaces, we define their **weak direct sum** to be

$$\coprod_{i \in I} A_i = \{f \mid f : I \longrightarrow \bigcup_{i \in I} A_i;\ f(i) \in A_i \text{ for each } i;\ \text{and } \mathrm{supp}(f) \text{ is finite}\}$$

considered as a subspace of $\prod_{i\in I} A_i$. Then $\coprod_{i\in I} A_i$ together with the injections

$$\text{in}_j : A_j \longrightarrow \coprod_{i\in I} A_i : a_j \mapsto \text{the } f \text{ with } f(i) = 0 \text{ for } i \neq j, \text{ and with } f(j) = a_j$$

is a coproduct of (A_i) in the category **Vect** in the sense of 1.

Proof: Consider the diagram

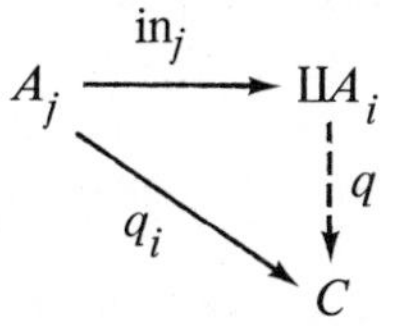

(3)

We define the map $q : \coprod_{i\in I} A_i \longrightarrow C : f \mapsto \sum_{i\in I} q_i(f(i))$ which is well-defined since supp(f) is finite. Then q is clearly linear, and is the one and only map for which (3) commutes. □

This relation of inclusion between $\coprod A_i$ and $\prod A_i$ in **Vect** is special. Contrast the situation in sets (Section 1.2).

In the category of the poset $(P, \leqslant)$, the diagram (1) says

$$a \leqslant a_i \text{ for all } i, \text{ while } c \leqslant a_i \;\Rightarrow\; c \leqslant a.$$

In other words, a product in $(P, \leqslant)$ is a **greatest lower bound** or **infimum**; and so arbitrary collections of objects in P have a product iff every subset of P has a greatest lower bound.

Dually, a coproduct is a **least upper bound**, or **supremum**.

For example, every collection of objects in the poset-category $(2^S, \subset)$ has both a product and a coproduct:

$$\coprod_{i\in I} A_i = \bigcup_{i\in I} A_i \quad \text{and} \quad \prod_{i\in I} A_i = \bigcap_{i\in I} A_i$$

so that **unions** and **intersections** have a pleasing categorical interpretation.

We next briefly discuss terminal and initial objects. The apparently simple ideas will prove very powerful in our study of limits in the remainder of the section.

5 **DEFINITION**: An object T in the category **K** is **terminal** iff for every object A of **K** there is a *unique* morphism $A \longrightarrow T$.

6 **OBSERVATION**: Terminal objects are unique up to isomorphism, for if T' is also terminal, there are unique morphisms $T \xrightarrow{\phi} T'$ and $T' \xrightarrow{\psi} T$ and we must have $\phi \cdot \psi = \text{id}_{T'}$ and $\psi \cdot \phi = \text{id}_T$ by the uniqueness of $T \longrightarrow T$ and $T' \longrightarrow T'$. □

Dually, we have

7 **DEFINITION**: An object I in the category **K** is **initial** iff for every object

A of **K** there is a unique morphism $I \longrightarrow A$.

In the category **Vect**, the vector space $\{0\}$ with only one element is both initial and terminal since $a \mapsto 0$ defines the only linear $A \longrightarrow \{0\}$, and $0 \mapsto 0$ defines the only linear $\{0\} \longrightarrow A$, for any vector space A.

In **Set**, the empty set $\emptyset$ is initial, while *any* one-element set is terminal.

In the category of the poset $(P, \leqslant)$ an element p is initial iff $p \leqslant p'$ for every p' in P; while p is terminal iff $p' \leqslant p$ for every p' in P. Thus P has an initial object iff it has a least element (usually denoted 0); and P has a terminal object iff it has a greatest element (usually denoted 1).

8 **DEFINITION**: An object 0 of a category **K** is called a **zero object** (compare the object $\{0\}$ of **Vect**) if it is both initial and terminal; i.e., for objects A and B, there is exactly one morphism $A \longrightarrow 0$ and $0 \longrightarrow B$.

Thus **Vect** has a zero object but **Set** does not.

9 **DEFINITION**: Given $f_1 : A_1 \longrightarrow A$ and $f_2 : A_2 \longrightarrow A$, we say the commutative diagram

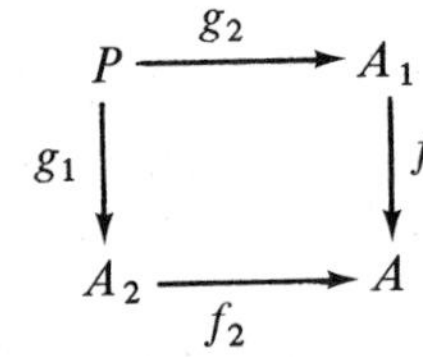

is a **pullback** (also (g_1, g_2) is a **pullback of** (f_1, f_2), g_1 is a **pullback of** f_1 **along** f_2, etc.), if it has the property that any commutative diagram $f_2 \cdot g_1' = f_1 \cdot g_2'$

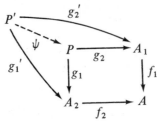

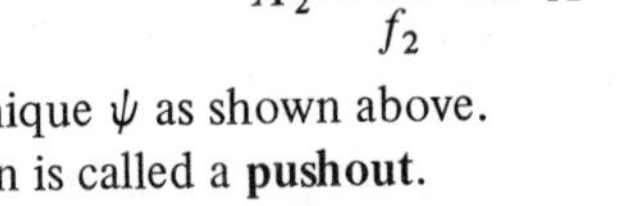

can be completed by a unique ψ as shown above.

The dual construction is called a **pushout**.

10 **DEFINITION**: Let **K** be a category with zero object 0. Then the **kernel** $u : K \longrightarrow A$ of $f : A \longrightarrow B$ is given by the pullback diagram

$$\begin{array}{ccc} K & \longrightarrow & 0 \\ {\scriptstyle u}\big\downarrow & & \big\downarrow \\ A & \underset{f}{\longrightarrow} & B \end{array}$$

See exercise 10.

We now provide a general framework for these concepts.

11 **DEFINITION**: A **directed graph** is an arbitrary class of **vertices**, together with an assignment, to each ordered pair (i, j) of vertices, of a class of **edges from** i **to** j. A directed graph is like a category without composition and identities; hence, we write $e : i \longrightarrow j$ to denote that e is an edge from i to j.

A **diagram** D in a category **K** is a directed graph whose vertices i are labelled by objects D_i of **K** and whose edges $i \longrightarrow j$ are labelled by morphisms in $\mathbf{K}(D_i, D_j)$.

A **cone** for a diagram D is a family $X \longrightarrow D_i$ of morphisms from a single object X such that $X \longrightarrow D_i \longrightarrow D_j = X \longrightarrow D_j$ for every $D_i \longrightarrow D_j$ in D. A morphism from a cone $(X \longrightarrow D_i)$ to a cone $(X' \longrightarrow D_i)$ is a **K**-morphism $X \longrightarrow X'$ such that $X \longrightarrow X' \longrightarrow D_i = X \longrightarrow D_i$ for all i. The cones for D then form a category, and a **limit for the diagram** D is a *terminal* object in this category, i.e., $(X \longrightarrow D_i)$ is such that for all cones $(X' \longrightarrow D_i)$ on D there is a *unique* morphism $X \longrightarrow X'$ such that $X' \longrightarrow X \longrightarrow D_i = X' \longrightarrow D_i$. We say that a limit $(X \longrightarrow D_i)$ for D has the **universal property** with respect to cones $(X' \longrightarrow D_i)$.

We say a family $(D_i \longrightarrow X)$ is a **colimit** for D if it is a limit for D considered in $\mathbf{K}^{\mathrm{op}}$. We say that a colimit has the **couniversal property** with respect to **cocones** $(D_i \longrightarrow X')$.

12 **OBSERVATION**: Since terminal objects are unique up to isomorphism, so too are limits and colimits.

13 **EXAMPLE**: In the poset category $(P, \leqslant)$, all diagrams commute. Hence the limit and colimit depend only on the vertices. Indeed, the limit of D (if it exists) is the infimum of the D_i, whereas the colimit is the supremum of the D_i.

14 **EXAMPLE**: Let A, B be objects of **K** and let D be the diagram

$$A \qquad\qquad B$$

(with no edges). Then the limit of D is the same concept as the product of A and B, whereas the colimit is the coproduct.

15 **EXAMPLE**: Consider the diagram D

$$\begin{array}{ccc} & & A_1 \\ & & \downarrow{\scriptstyle f_1} \\ A_2 & \xrightarrow[f_2]{} & A \end{array}$$

with three vertices and two edges. Strictly speaking, a cone for D is an object C and three morphisms g_1, g, g_2 :

$$\begin{array}{ccc} C & \xrightarrow{g_1} & A_1 \\ {\scriptstyle g_2}\downarrow & \overset{g}{\searrow} & \downarrow{\scriptstyle f_1} \\ A_2 & \xrightarrow[f_2]{} & A \end{array}$$

But this is equivalent to

$$\begin{array}{ccc} C & \xrightarrow{g_1} & A_1 \\ {\scriptstyle g_2}\downarrow & & \downarrow{\scriptstyle f_1} \\ A_2 & \xrightarrow[f_2]{} & A \end{array}$$

since g is determined as the common morphism $f_2 \cdot g_2 = f_1 \cdot g_1$ from C to A. Hence the limit of D is the pullback of (f_1, f_2).

Again, consider the diagram D

$$A \underset{f_2}{\overset{f_1}{\rightrightarrows}} B$$

Strictly speaking, a cone for D comprises both an $h : C \longrightarrow A$ and a $g : C \longrightarrow B$, but g can be omitted since it must be the common value $f_1 \cdot h = f_2 \cdot h$. Thus a limit of D is an equalizer of f_1 and f_2.

16 **DEFINITION**: We say a category **has (finite) limits** if every (finite) diagram has a limit. (A diagram is finite if its sets of edges and vertices are both finite.)

So far, we have not considered to what extent limits and colimits are expected to exist, even in nice categories such as **Set**, **Vect** and **Poset**. The following remarkable theorem shows that if simple diagrams have limits, complicated ones do too. (For the concept of a *small* diagram or limit, see exercise 13.)

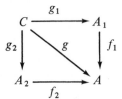

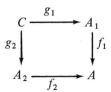

17 **THEOREM**: Let D be a diagram in **K** with sets V of vertices and E of edges. Then if every V-indexed family and every E-indexed family of objects in **K** has a product and if every pair of **K**-morphisms (between the same two objects) has an equalizer, D has a limit.

Proof: Form the V-indexed and E-indexed products shown below. The universal property of the E-indexed product induces unique ψ_1 and ψ_2 as shown.

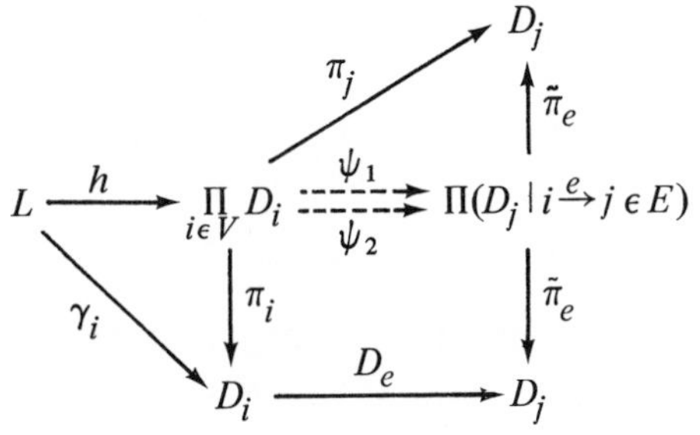

(4)

Let $h = \mathrm{eq}(\psi_1, \psi_2)$. Set $\gamma_i = \pi_i \cdot h$ as shown above. We claim that $(L, (\gamma_i))$ is a cone over D. If $e : i \longrightarrow j$ is in E, we must show

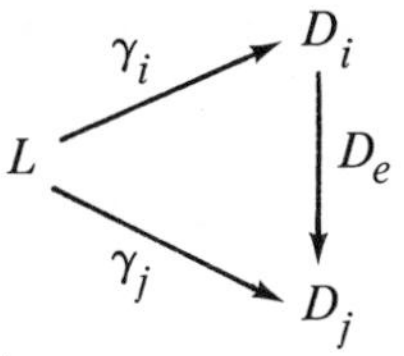

(5)

But

$$D_e \cdot \gamma_i = D_e \cdot \pi_i \cdot h = \tilde{\pi}_e \cdot \psi_2 \cdot h = \tilde{\pi}_e \cdot \psi_1 \cdot h = \pi_j \cdot h = \gamma_j .$$

Now suppose that $(L', (\gamma_i'))$ is an arbitrary cone over D. By the universal property of the V-indexed product, there exists unique h' with $\pi_i \cdot h' = \gamma'_i$. (Look at (4) imagining primes where necessary.) Consulting (5), substituting primes where necessary, we have

$$\tilde{\pi}_e \cdot \psi_1 \cdot h' = \pi_j \cdot h' = \gamma'_j = D_e \cdot \gamma'_i = D_e \cdot \pi_i \cdot h' = \tilde{\pi}_e \cdot \psi_2 \cdot h' .$$

As e is arbitrary, $\psi_1 \cdot h' = \psi_2 \cdot h'$. Since $h = \mathrm{eq}(\psi_1, \psi_2)$ there exists unique $\Gamma : L' \longrightarrow L$ such that $h \cdot \Gamma = h'$.

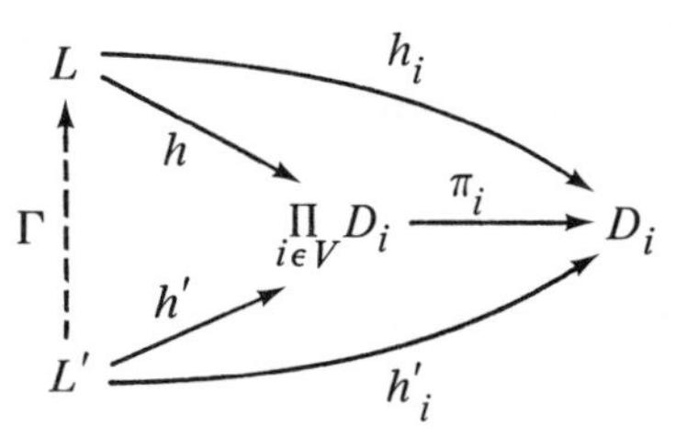

Thus

$$h_i \cdot \Gamma = \pi_i \cdot h \cdot \Gamma = \pi_i \cdot h' = h'_i \, .$$

If also $\tilde{\Gamma}$ satisfies $h_i \cdot \tilde{\Gamma} = h'_i$ then as $\pi_i \cdot h \cdot \tilde{\Gamma} = \pi_i \cdot h'$ for all i, $\tilde{\Gamma} \cdot h = h'$ and so $\Gamma = \tilde{\Gamma}$. □

18 **COROLLARY**: The limit of any diagram D in **Set** is obtained as follows. A function $f : V \longrightarrow \cup D_i$ with $f(i) \in D_i$ – that is, a typical element of ΠD_i – can be written in 'tuple' notation (d_i); thus $d_i = f(i)$ and (d_i) is the function $i \mapsto d_i$. Let L be the subset of $\Pi(D_i \mid i \text{ is a vertex})$ of all tuples (d_i) such that for all edges $e : i \longrightarrow j$, $D_e(d_i) = d_j$. Let $\gamma_i : L \longrightarrow D_j$ be the restricted projection, $(d_i) \mapsto d_j$. Then $(L, (\gamma_i))$ is the limit of D.

Proof: Apply the construction in the proof of 17 in the context of **Set**. □

19 **COROLLARY**: The colimit of any diagram D in **Set** is obtained as follows. Consider the disjoint union $\amalg D_i = \{(d, i) \mid d \in D_i\}$. Let R be the subset

$$\{(d, i; d', i') \mid \text{there exists } e : i \longrightarrow i' \text{ with } D_e(d) = d'\}$$

of $\amalg D_i \times \amalg D_i$ and let $\bar{R}$ be the equivalence relation generated by R as in 1.3.8. Let $\gamma_i : D_j \longrightarrow (\amalg D_i)/R$, $d \mapsto [d, j]$. Then $((\amalg D_i)/R, (\gamma_j))$ is the colimit of D.

Proof: Apply the construction in the dualization of the proof of 17 to the construction of coproducts and coequalizers of pairs in **Set**. □

Exercises

1 Let **K** be a category in which the product $A \times B$ exists for all A, B. Prove that $(A \times B) \times C \cong A \times (B \times C)$. If **K** has a terminal object I, prove $A \times I \cong A \cong I \times A$.

2 Let D be the empty diagram with no vertices and hence no edges. Show that a limit of D is a terminal object whereas a colimit of D is an initial object.

3 Show that essentially the same construction as in 18 constructs limits of diagrams in **Vect** and **Poset**.

4 Use exercise 3.2 to conclude that the analogous construction to 19 does not construct colimits of diagrams in **Poset**.

5 Discover the analog of 19 for **Vect**.

6 Use 17 to discover how pullbacks can be constructed from products and equalizers.

7 Let $\mathbf{S}(B)$ be the class of mono-subobjects of B (3.9). If $[A, f]$, $[A', f']$ $\in \mathbf{S}(B)$ say that $[A, f] \leqslant [A', f']$ if there exists

$$\begin{array}{ccc} A & \xrightarrow{f} & B \\ \downarrow{\scriptstyle \text{(dashed)}} & \nearrow_{f'} & \\ A' & & \end{array}$$

Show that this is well-defined (i.e. depends only on the equivalence classes, not the particular monomorphisms f, f') and defines a partial order on $\mathbf{S}(B)$. Prove that the limit

$$\begin{array}{ccc} A & \xrightarrow{f} & B \\ {\scriptstyle \gamma_i}\downarrow & & \\ A_i & & \end{array} \qquad \text{of the diagram} \qquad \begin{array}{ccc} & & B \\ & \nearrow^{f_i} & \\ A_i & & \end{array}$$

with $[A_i, f_i] \in \mathbf{S}(B)$ is such that f is a monomorphism and show that $[A, f]$ is the infimum of $[A_i, f_i]$ in the poset $(\mathbf{S}(B), \leqslant)$. Conversely, show that if such an $[A, f]$ is the infimum of $[A_i, f_i]$ – so that $[A, f] \leqslant [A_i, f_i]$ via $\gamma_i : A \longrightarrow A_i$ in particular – then $f : A \longrightarrow B$ and $\gamma_i : A \longrightarrow A_i$ provide a limit for $f_i : A_i \longrightarrow B$. For these reasons, limits of diagrams $f_i : A_i \longrightarrow B$ with each f_i a monomorphism are called **intersections.** Notice, however, that the dual concept is *not* 'unions' of subobjects.

8 Prove that the pullback of a monomorphism is a monomorphism; i.e., in 9, if f_1 is mono prove g_1 is mono. Similarly, prove that the pullback of an equalizer is an equalizer and that the pullback of a split epimorphism is a split epimorphism. Give an example in **Set** to show that the pullback of a split monomorphism need not be a split monomorphism.

9 Consider the commutative diagram

$$\begin{array}{ccccc} A_1 & \xrightarrow{f_1} & B_1 & \xrightarrow{g_1} & C_1 \\ {\scriptstyle a}\downarrow & & {\scriptstyle b}\downarrow & & \downarrow{\scriptstyle c} \\ A_2 & \xrightarrow[f_2]{} & B_2 & \xrightarrow[g_2]{} & C_2 \end{array}$$

Prove that if both squares are pullbacks then the outer rectangle is a pullback.

10 Let $f : X \longrightarrow Y$ be a linear map in **Vect**. Prove that the kernel of f as in 10 is the usual construction of the subspace $\{x \in X \mid f(x) = 0\}$. Also prove that the pushout

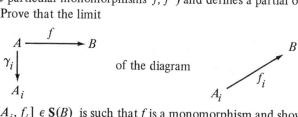

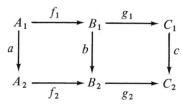

$$\begin{array}{ccc} A & \xrightarrow{f} & B \\ \downarrow & & \downarrow{\scriptstyle \eta} \\ 0 & \longrightarrow & C \end{array}$$

describing the cokernel of f is the construction $\eta : B \longrightarrow B/f(A)$ used in the proof of 3.11.

11 Let (**E**, **M**) be an image factorization system. Prove that the pullback of a morphism in **M** is again in **M**. [Hint: Use exercise 3.12.]

12 Let **K** be the 4-object 9-morphism category

$$A \underset{b}{\overset{a}{\rightrightarrows}} B \xrightarrow{c} C \xleftarrow{d} D$$

i.e. the morphisms are $\mathrm{id}_A, \mathrm{id}_B, \mathrm{id}_C, \mathrm{id}_D, a, b, c, d$ and $c \cdot a = c \cdot b$. Show that $C = B + D$ but that the coproduct injection c is not a monomorphism. Show that whenever $A = \amalg A_i$ in a category **L**, a sufficient condition for $\mathrm{in}_j : A_j \longrightarrow A$ to be a split monomorphism is that there exists a morphism from A_j to A_i for all i.

13 Show that the product P of all non-empty sets X in **Set** is not itself in **Set**. [Hint: If P is in **Set** so is the power set 2^P. Thus we have the projection $\pi : P \longrightarrow 2^P$ of P to the set 2^P, which is clearly a surjection. Define A in 2^P by $A = \{x \in P \mid x \notin \pi(x)\}$ and let $\pi(y) = A$. Then "$y \in A$" implies "$y \notin A$" whereas "$y \notin A$" implies "$y \in A$", the desired contradiction.] Discussion: Thus there exist collections which are not sets. We say a collection is **small** if it is a set, i.e. an object of **Set**. We say a *diagram* is **small** just in case its collection of vertices forms a set, and its collection of edges forms a set. Thus we may paraphrase Theorem 17 as saying that if a category **K** has equalizers and small products, then it has small limits. For more on this subtle point, see Mac Lane's "Categories for the Working Mathematician", Section I.6.

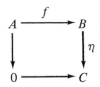

Chapter 3

MONOIDS AND GROUPS

In this short chapter we exemplify the theory of Chapter 2 in some interesting categories of algebraic structures.

3.1 DEFINING THE CATEGORIES

We have already seen (2.1.1) that a vector space is, in part, a set X equipped with a binary operation $+ : X \times X \longrightarrow X$ and a distinguished element $0 \in X$ satisfying (amongst other equations)

$$x + (y + z) = (x + y) + z \qquad \text{for all } x, y, z \in X$$

$$x + 0 = x = 0 + x \qquad \text{for all } x \in X$$

For all $x \in X$ there exists $-x \in X$ such that

$$x + (-x) = 0 = (-x) + x\,.$$

The same type of structure can arise in other ways. For example, consider the set $S = \{a, b, c\}$ and let X be the set of all **permutations** of S (i.e., bijections, isomorphisms, one-to-one-onto mappings from S to itself). Then X has exactly six elements as described below:

$$\begin{array}{ll}
\mathrm{id} : S \longrightarrow S & \mathrm{id}(a) = a,\ \mathrm{id}(b) = b,\ \mathrm{id}(c) = c \\
\sigma_a : S \longrightarrow S & \sigma_a(a) = a,\ \sigma_a(b) = c,\ \sigma_a(c) = b \\
\sigma_b : S \longrightarrow S & \sigma_b(a) = c,\ \sigma_b(b) = b,\ \sigma_b(c) = a \\
\sigma_c : S \longrightarrow S & \sigma_c(a) = b,\ \sigma_c(b) = a,\ \sigma_c(c) = c \\
\alpha : S \longrightarrow S & \alpha(a) = b,\ \alpha(b) = c,\ \alpha(c) = a \\
\beta : S \longrightarrow S & \beta(a) = c,\ \beta(b) = a,\ \beta(c) = b
\end{array}$$

Functional composition $X \times X \longrightarrow X : (x, y) \mapsto x \cdot y$ is a binary operation on X. We have

$$x \cdot (y \cdot z) = (x \cdot y) \cdot z$$

$$x \cdot \mathrm{id} = x = \mathrm{id} \cdot x$$

$$x \cdot x^{-1} = \mathrm{id} = x^{-1} \cdot x$$

where x^{-1} denotes the inverse function $x(s) \mapsto s$ to the function x (e.g.

$\sigma_a^{-1} = \sigma_a$, $\alpha^{-1} = \beta$, etc.). This indeed looks like the vector space situation if we substitute '$\cdot$' for '+', 'id' for '0' and 'x^{-1}' for '$-x$'. Notice that the law $x + y = y + x$ for vector space addition does not carry over to $x \cdot y$, since $\sigma_a \cdot \alpha = \sigma_b$ but $\alpha \cdot \sigma_a = \sigma_c$. We formalize with

1 **DEFINITION**: A **group** is a set X equipped with a function $\cdot : X \times X \to X$ (the group **multiplication**) and a distinguished element $e \in X$ (the group **identity**) subject to the three laws

(associativity)	$x \cdot (y \cdot z) = (x \cdot y) \cdot z$ for all $x, y, z \in X$
(e is a two-sided identity)	$e \cdot x = x = x \cdot e$ for all $x \in X$
(existence of **inverses**)	for all $x \in X$ there exists $x^{-1} \in X$ such that $x \cdot x^{-1} = e$ and $x^{-1} \cdot x = e$.

For another example of a group, let X be the set of all real numbers > 0, let $x \cdot y$ be ordinary multiplication and let $e = 1$, $x^{-1} = \frac{1}{x}$.

Certain mathematical systems satisfy the first two group axioms without satisfying the third. For example, if X is the set of all real numbers with $x \cdot y$ denoting ordinary multiplication and $e = 1$ then $x \cdot (y \cdot z) = (x \cdot y) \cdot z$ and $x \cdot e = x = e \cdot x$ but there is no x^{-1} with $x \cdot x^{-1} = e$ if $x = 0$. For another example, let S be the set $\{a, b, c\}$ as before, but now consider the set Y of all functions $S \longrightarrow S$. Then Y has 27 elements as opposed to the 6 permutations. With $x \cdot y =$ functional composition and $e = \text{id}$, we again have $(x \cdot y) \cdot z = x \cdot (y \cdot z)$ and $x \cdot e = x = e \cdot x$. However, if $\tilde{a} \in Y$ is the constant function $\tilde{a}(a) = \tilde{a}(b) = \tilde{a}(c) = a$ there is no $(\tilde{a})^{-1}$ with $(\tilde{a})^{-1} \cdot \tilde{a} = e$. We formalize with

2 **DEFINITION**: A **monoid** is a set X equipped with a function $\cdot : X \times X \longrightarrow X$ (the monoid **multiplication**) and a distinguished element (the monoid **identity**) subject to the two laws

$$x \cdot (y \cdot z) = (x \cdot y) \cdot z \qquad \text{for all } x, y, z$$

$$x \cdot e = x = e \cdot x \qquad \text{for all } x.$$

We shall often abbreviate $x \cdot y$ to xy, etc.

To dispel the idea that monoids arise by adding extraneous elements to groups, consider the classic example that the set **N** of natural numbers $\{0, 1, 2, 3, ...\}$ forms a monoid under addition with identity 0.

We say X' is a **sub-monoid** of the monoid X if X' contains the identity and is closed under multiplication: if x_1, x_2 in X' then $x_1 x_2$ in X'. Similarly, G' is a **subgroup** of X if it is a submonoid of X on which the multiplication admits inverses: to each x in G' there corresponds an x^{-1} in G' with $xx^{-1} = e = x^{-1}x$. Note that a submonoid of a group need not be a group: **N** is a submonoid of the group **Z** of all integers under addition, but is not itself a group.

An example that is at the heart of monoid theory also plays a central role in the *theory of computation*: consider a set A (the 'alphabet') and let X be the set of all *words on the alphabet* A, i.e., X is the set of all strings $a_1 \ldots a_n$ with each $a_i \in A$. We do not exclude the case $n = 0$; there is one such string, call it Λ, the 'empty string'. Define $(a_1 \ldots a_n) \cdot (b_1 \ldots b_m) = a_1 \ldots a_n b_1 \ldots b_m$ and define $e = \Lambda$. With this structure, X becomes a monoid. X is denoted by A^*, and is called the **free monoid on the set A of generators**, for reasons that will become clear in Chapter 7.

If **K** is any category and A is an object of **K**, the set $X = \mathbf{K}(A, A)$ is a monoid with $x \cdot y$ the category composition and $e = \mathrm{id}_A$. Note that this generalizes our previous example of $\mathbf{K} = \mathbf{Set}$, $A = \{a, b, c\}$.

If **K** is a category with only one object, A, the data and axioms defining **K** coincide with the data and axioms by virtue of which $\mathbf{K}(A, A)$ is a monoid. We deduce:

3 A monoid is the same thing as a category with one object. □

If X, Y are monoids, a function $f: X \longrightarrow Y$ is a monoid **homomorphism** if f preserves the monoid structure of multiplication and identities, i.e., if $f(x \cdot x') = f(x) \cdot f(x')$ and $f(e) = e$.

The **identity function** $\mathrm{id}_X : X \longrightarrow X$ is surely a monoid homomorphism and the usual **composite** $g \cdot f : X \longrightarrow Z$ of two functions is a monoid homomorphism if $f: X \longrightarrow Y$ and $g: Y \longrightarrow Z$ are. Thus:

4 Monoids and their homomorphisms form a category **Mon**. □

Groups form a category if we define a **group homomorphism** to be a monoid homomorphism from one group to another (noting that every group may be considered as a monoid if we *forget* about inversion). Given the observation that a submonoid of a group need not be a subgroup, it might be questioned, *a priori*, whether or not the group homomorphism $f: X \longrightarrow Y$ preserves the additional structure of inversion – that is, whether $f(x^{-1}) = (f(x))^{-1}$ – but this is true automatically because $f(x)f(x^{-1}) = f(xx^{-1}) = f(e) = e$ and $f(x^{-1})f(x) = e$ similarly, so that $f(x)$ and $f(x^{-1})$ are indeed inverses.

5 Groups and their homomorphisms form a category **Grp**. □

Exercises

1 Prove that the unit e is unique in any monoid. Then prove that the two-sided inverse x^{-1} of an element x is unique if it exists at all.

2 Let A be a set and let $X = 2^A$. Show that X is a group via $S_1 S_2 = \{a \mid a \in S_1 \text{ or } a \in S_2 \text{ but } a \notin S_1 \cap S_2\}$ with identity $\emptyset$. Why is $S_1 S_2 = S_1 \cup S_2$ *not* a group structure?

3 Let X, Y, Z be monoids and let $f : X \longrightarrow Y$, $g : Y \longrightarrow Z$ be functions such that f is onto and f, $g \cdot f$ are monoid homomorphisms. Prove that g is a monoid homomorphism.

4 Prove that $\{0, 1\}$ is a group via $x + 0 = x = 0 + x$, $1 + 1 = 0$. This group is known as $\mathbf{Z}_2$. Prove that $P : \mathbf{Z} \longrightarrow \mathbf{Z}_2$, where p is the *parity map* $p(n) = 0$ if n is even, $= 1$ if n is odd is a group homomorphism. Prove that p is *not* a retraction in **Grp**.

5 Let A be an object in a category **K**. Show that the subset of $\mathbf{K}(A, A)$ of **K**-isomorphisms is a group.

6 Let M be a monoid. Show that 2^M is a monoid under 'setwise' multiplication $S_1 S_2 = \{xy \mid x \in S_1 \text{ and } y \in S_2\}$. Is 2^M a group if M is a group?

7 Construct an example of a monoid M with an element $u \neq e$ such that $\{u\}$ is a submonoid of M (but not a subgroup, even though the monoid $\{u\}$ is a group).

8 Let S be a set and let A be a subset of S. Let X be the set of all bijections $f : S \longrightarrow S$ for which $\{f(a) \mid a \in A\} = A$. Show that X is a group under functional composition.

9 Let $f : X \longrightarrow Y$ be a bijective monoid homomorphism. Prove that $f^{-1} : Y \longrightarrow X$ is a monoid homomorphism (and, hence, that f is a monoid isomorphism).

3.2 CONSTRUCTIONS WITHIN THE CATEGORIES

1 *Products in* **Mon** *and* **Grp**: It is easy to see that any family $(X_i \mid i \in I)$ of monoids has a product. Set X to be the usual cartesian product set with coordinate projections $p_i : X \longrightarrow X_i$. A typical element of X may be written (x_i) (see 2.4.18) so that $p_j((x_i)) = x_j$ for each $j \in I$. Then there exists at most one multiplication and identity on X rendering each p_i a monoid homomorphism, namely the componentwise multiplication $(x_i)(y_i) = (x_i y_i)$, and the 'string of identities' (e_i). That this is a monoid structure on X is clear. It only remains to note that the diagram

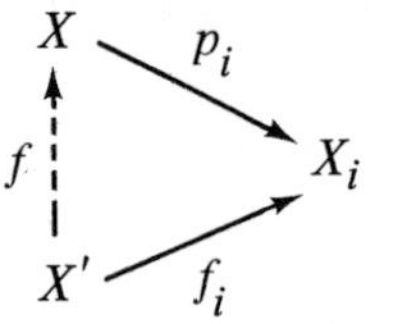

is uniquely completed for all i by the assignment

$$f(x') = (y_i) \quad \text{where each } y_i = f_i(x')$$

and that this f is indeed a homomorphism:

$$f(e') = (e_i)$$

and $\quad f(xx') = (f_i(xx')) = (f_i(x) \cdot f_i(x')) = (f_i(x)) \cdot (f_i(x')) = f(x) \cdot f(x')$.

Thus the family $(p_i : X \longrightarrow X_i \mid i \in I)$ is indeed the product of the monoids X_i.

Moreover, X is a group if each X_i is, so is *a fortiori* the product in **Grp**, because of the special property that any monoid homomorphism between groups is a group homomorphism.

2 *Equalizers of pairs in* **Mon** and **Grp**: Let $f, g : X \longrightarrow Y$ be monoid homomorphisms. Set

$$E = \{x \in X \mid f(x) = g(x)\}$$

and let $i : E \longrightarrow X$ be the inclusion function $i(x) = x$. E is closed under the multiplication of X (i.e., if $x, y \in E$ then $f(xy) = f(x)f(y) = g(x)g(y) = g(xy)$, so $xy \in E$) and contains the identity (as $f(e) = e = g(e)$). This may be summed up by saying there exists a unique monoid structure on E such that $i : E \longrightarrow X$ is a homomorphism. The definition ensures that $f \cdot i = g \cdot i$. It is now easy to verify the universal property that makes $i : E \longrightarrow X$ an equalizer:

Let $h : Z \longrightarrow X$ be a monoid homomorphism satisfying $f \cdot h = g \cdot h$.

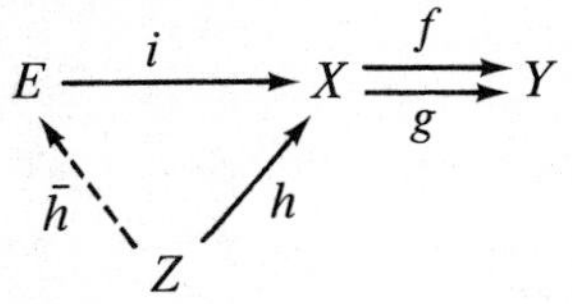

Then $h(Z) \subset E$ so that $\bar{h} : Z \longrightarrow E$ defined by $\bar{h}(z) = h(z)$ is the unique function with $i \cdot \bar{h} = h$. Clearly, $\bar{h}$ is a monoid homomorphism because h is. Therefore $i = \mathrm{eq}(f, g)$ in **Mon**. If X and Y are groups, so is E (because if $f(x) = g(x)$ then $f(x^{-1}) = f(x)^{-1} = g(x)^{-1} = g(x^{-1})$, cf. the discussion before 1.5) and $i = \mathrm{eq}(f, g)$ in **Grp**.

3 *Coproducts in* **Mon** *and* **Grp**: The one element monoid/group $\{e\}$ is clearly the initial object – that is, the empty coproduct – in both categories.

Let now $\{X_i \mid i \in I\}$ be a family of monoids with I non-empty. Let A (for 'alphabet') be the disjoint union of the sets X_i. Let W be the set A^* of all words on the alphabet A, as in the previous section. A word $(a_1, \ldots, a_n)$ is *reduced* if for all $j < n$, $i_j \neq i_{j+1}$ and for all $j \leqslant n$, $x_j \neq e_j$ (where $a_j = (x_j, i_j)$ and e_j is the unit of X_{i_j}). If a word is not reduced it has a *reduction* obtained by continuing to apply the following rules:

(a) Multiply adjacent pairs with the same i-term:

$$(a_1, \ldots, a_j = (x_j, i), a_{j+} = (x_{j+1}, i), \ldots, a_n) \nleftrightarrow (a_1, \ldots, \hat{a}_j = (x_j x_{j+1}, i), a_{j+2}, \ldots, a_n)$$

(b) Delete identity terms:

$$(a_1, \ldots, a_j = (e_j, i_j), \ldots, a_n) \nleftrightarrow (a_1, \ldots, a_{j-1}, a_{j+1}, \ldots, a_n)$$

until the word is reduced. It is obvious that such a process has at most n steps and results in a reduced word. The monoid axioms guarantee that the end result is independent of the way the rules are applied (why?). Let C be the set of all reduced words in W. Define a multiplication on C by

$$(a_1, \ldots, a_n)(b_1, \ldots b_m) = \text{reduction of } (a_1, \ldots a_n, b_1, \ldots, b_m).$$

Then C is a monoid with unit Λ. Define injection homomorphisms

$$\text{in}_i : X_i \longrightarrow C \qquad x_i \mapsto \text{reduction of } (x_i, i)$$

These *are* monoid homomorphisms by the reduction rules. To check that $(\text{in}_i : X_i \longrightarrow C \mid i \in I)$ is indeed a coproduct for the monoids X_i we must check that whenever $f_i : X_i \longrightarrow D$ are monoid homomorphisms, the diagram

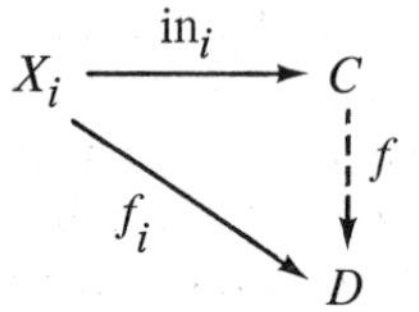

commutes for a unique monoid homomorphism f. But for f to be a homomorphism it clearly must satisfy $f(\Lambda) = e$, while, for $n > 0$, $f(a_1, \ldots, a_n) = f_{i_1}(x_1)\ldots f_{i_n}(x_n)$ where $a_j = (x_j, i_j)$. But this determines f as a function. The proof that this f is a monoid homomorphism is left as a routine exercise.

It is true that C is a group if each X_i is. The coproduct in **Grp** is, therefore, constructed in the same way.

Now we turn to coequalizers. Let Y be a monoid. A **monoid congruence** on Y is a subset $R \subset Y \times Y$ which is simultaneously an equivalence relation and a submonoid of $Y \times Y$ (i.e., if $(a, x) \in R$ and $(x, b) \in R$ then $(ab, xy) \in R$). In this case there exists a unique monoid structure on Y/R admitting $\eta_R : Y \rightarrow Y/R$ as a monoid homomorphism, namely

$$[a]\,[b] = [ab], \qquad \text{and } [e] \text{ as identity.}$$

Uniqueness is clear. For existence we must prove that $[ab] = [xy]$ if $[a] = [x]$ and $[b] = [y]$ (that is, we must ensure that the definition of multiplication is well-defined); but this is just our assumption that R is a submonoid of $Y \times Y$.

Suppose $\{R_i \mid i \in I\}$ is a family of monoid congruences on Y. Then

$$R = \cap R_i = \{(x, y) \mid (x, y) \in R_i \text{ for all } i \in I\}$$
$$(= Y \times Y \text{ if } I \text{ is empty})$$

is obviously a monoid congruence on Y.

Let $h : Y \longrightarrow Z$ be a monoid homomorphism. Then

$$\{(y_1, y_2) \mid h(y_1) = h(y_2)\}$$

is clearly a monoid congruence on Y.

4 *Coequalizers in* **Mon** *and* **Grp**: Let $f, g : X \longrightarrow Y$ be monoid homomorphisms and define $R_0 \subset Y \times Y$ by

$$R_0 = \{(f(x), g(x)) \mid x \in X\}.$$

and set

$$\bar{R} = \cap \{R \mid R \text{ is a monoid congruence on } Y,\ R_0 \subset R\}.$$

Then $\eta_R : Y \longrightarrow Y/\bar{R} = \text{coeq}(f, g)$.

Proof: Let $h : Y \longrightarrow Z$ be a monoid homomorphism such that $h \cdot f = h \cdot g$.

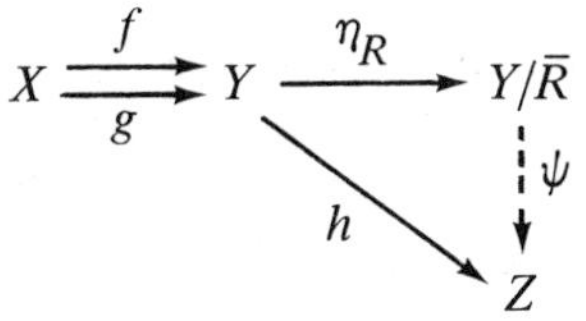

Clearly $R_0 \subset R = \{(y_1, y_2) \mid h(y_1) = h(y_2)\}$. As R is a monoid congruence, $\bar{R} \subset R$. Therefore $\psi[y] = h(y)$ is well-defined, is easily checked to be a monoid homomorphism and is surely unique with these properties.

If X and Y are groups (or, in fact, if just Y is a group) then $Y/\bar{R}$ is a group since $[y][y^{-1}] = [e] = [y^{-1}][y]$. The same construction, then, provides coequalizers in **Grp**.

5 *Monomorphisms in* **Mon** *and* **Grp**: It is obvious that if $f : X \longrightarrow Y$ is a monoid homomorphism which is one-to-one then it is a monomorphism in **Mon.** Let us now prove the converse: suppose $f : X \longrightarrow Y$ is a monomorphism in **Mon** and suppose $x \neq y \in X$. Let **N** be the monoid of natural numbers $\{0, 1, 2, ...\}$ under addition and define, for each $z \in X$, the monoid homomorphism

$$\tilde{z} : \mathbf{N} \longrightarrow X : n \mapsto z^n$$

where the scheme

$$z^0 = e$$
$$z^{n+1} = (z^n)z$$

defines z^n. To prove $\tilde{z}$ is a homomorphism is to prove $\tilde{z}(n) \cdot \tilde{z}(m) = \tilde{z}(n + m)$, i.e., $z^n z^m = z^{n+m}$ which is easy. Since $\tilde{x}(1) = x \neq y = \tilde{y}(1)$, $\tilde{x} \neq \tilde{y}$, and $f \cdot \tilde{x} \neq f \cdot \tilde{y}$ since f is a monomorphism. If $g : \mathbf{N} \longrightarrow Y$ is any homomorphism, $g(n) = (g(1))^n$ and so g is determined by $g(1)$. Thus $f \cdot \tilde{x}(1) \neq f \cdot \tilde{y}(1)$ and so $f(x) = (f \cdot \tilde{x})(1) \neq (f \cdot \tilde{y})(1) = f(y)$, and f is one-to-one. Thus, monomorphisms in **Mon** coincide with one-to-one homomorphisms. The same statement is true in **Grp**; the argument is the same, replacing **N** with the group **Z** of (positive and negative) integers.

6 *Epimorphisms in* **Mon** *and* **Grp**: The inclusion homomorphism $i : \mathbf{N} \longrightarrow \mathbf{Z}$ is a morphism in **Mon** which is one-to-one, and is therefore a monomorphism. It

is also an epimorphism! For let $f, g : \mathbf{Z} \longrightarrow X$ be monoid homomorphisms satisfying $f \cdot i = g \cdot i$. Then for all $n \in \mathbf{N}$ we have $f(n)f(-n) = f(0) =$ $= g(n)g(-n) = f(n)g(-n)$ so that $f(-n) = f(n)^{-1} = g(-n)$, and $f = g$.

Thus *epimorphisms need not be onto in* **Mon** and morphisms in **Mon** which are simultaneously epimorphic and monomorphic need not be isomorphisms. On the other hand, a homomorphism in **Grp** is epimorphic if and only if it is onto (see exercise 6).

Exercises

1 A monoid X is **abelian** if $xy = yx$ for all x, y. Let **Abm** be the category of abelian monoids and monoid homomorphisms. Prove that products, equalizers and coequalizers can be constructed by mimicking these constructs in **Mon** or **Grp**. Construct coproducts in **Abm** by imitating the construction in **Vect**.

2 Let X be the two-element group $\{e, a\}$ $(aa = e)$. Prove that $X + X$ is infinite.

3 Prove that a homomorphism in **Mon** or **Grp** is a coequalizer if and only if it is onto.

4 Let Y be a monoid, let $R_0 \subset Y \times Y$ and let $\bar{R}$ be the intersection of all monoid congruences containing R_0 as in 4. Let $\tilde{R}$ be the equivalence relation generated by R_0 as in 1.3.8. Prove that if R_0 is a submonoid and if $(y, y) \in R_0$ for all $y \in Y$ then $\bar{R} = \tilde{R}$.

5 Let A have one element. Prove that the free monoid A^* is isomorphic to **N** in the category of monoids. If A has n elements, is it then true that $A^* \cong \mathbf{N}^n$?

6 Prove that if $\phi : G \longrightarrow H$ is a group homomorphism with $\phi(G) = M \neq H$, then ϕ is not an epimorphism in **Grp**. [Only attempt the exercise if you know enough group theory to understand the Hint: If M has only 2 cosets in H, take $K = H/M$, and define $\psi, \psi' : H \longrightarrow K$ by $\psi(h) = [h]$ and $\psi'(h) = e_K$. If not, take K to be the group of all permutations of H. Choose 3 different cosets M, Mh', Mh'' of M in H, and define σ in K by $\sigma(xh'') = xh'$, $\sigma(xh') = xh''$ for $x \in M$, while otherwise $\sigma(h) = h$. Define $\psi, \psi' : H \longrightarrow K$ by $\psi(h) =$ left multiplication by h, and $\psi'(h) = \sigma^{-1}\psi(h)\sigma$. Then in either case $\psi\phi = \psi'\phi$ but $\psi \neq \psi'$.]

7 Prove that (onto, one-to-one) is an image factorization system both in **Grp** and in **Mon**.

8 A **semilattice** is a poset in which every finite subset has a least upper bound. Show that a semilattice is the same thing as an abelian monoid satisfying $x \cdot x = x$ for all x. [Hint: $LUB\ \{x_1, \ldots, x_n\} = x_1 + \ldots + x_n$ $(= e$ if $n = 0)$

whereas $x \leqslant y$ iff $xy = y$.] Verify that a monoid homomorphism between semilattices preserves all finite LUB's and is automatically order-preserving. Construct an order-preserving map from the semilattice **R** (with the usual ordering) to itself which is not a monoid homomorphism. [Hint: Any monotone increasing function satisfies the former whereas the latter must be a straight line.]

9 Verify that a product of semilattices, as constructed in **Abm**, is still a semilattice. State and prove a similar statement for equalizers, coequalizers and coproducts.

10 Show that, replacing (abelian monoids, semilattices) with (monoids, abelian monoids), exercise 9 goes through for products, equalizers and coequalizers but not for binary coproducts.

Chapter 4

METRIC AND TOPOLOGICAL SPACES

We examine categorical constructions in some 'non-algebraic' categories.

4.1 CATEGORIES OF METRIC SPACES

It is tempting to regard the distance between two points in the real world as completely determined by a choice of scale, but a little thought suggests manifold ways of measuring distance. Consider the triangle ABC with sides a, b, c shown below. It is not necessarily true that the distance from A to C should be defined to be c.

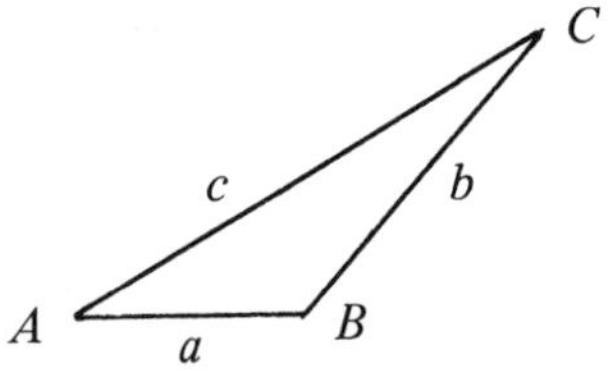

In a city with streets running parallel to AB and BC, the only practical route from A to C might be A-B-C and $a + b$ is the 'true' distance. Then again, the distance from New York to Barcelona is not likely to be measured on a true straight line (which goes through the Earth) but along a great circle on the Earth's surface.

The mathematical approach to distance is to axiomatize a few properties that all notions of distance share.

1 **DEFINITION**: A **metric** (or **distance function**) **on** a set X is a function $d : X \times X \longrightarrow \mathbf{R}_+$ (where $\mathbf{R}_+$ is the set of all real numbers $\geqslant 0$) which satisfies the following three axioms:

1. For all $x, y \in X$, $d(x, y) = 0$ if and only if $x = y$.
2. For all $x, y \in X$, $d(x, y) = d(y, x)$.
3. For all $x, y, z \in X$, $d(x, y) + d(y, z) \geqslant d(x, z)$.

The heuristic content of the first two axioms is clear: the distance from a point to itself is 0; the distance between distinct points is > 0; and the distance from x to y is the same as the distance from y to x. The third axiom – called the **triangle inequality** – asserts that one cannot shrink the total distance travelled by making detours. For example, let $X = \mathbf{R}$ be the set of real numbers and define $d(x, y) = |x - y|$. Then $|x - y| = 0$ if and only if $x = y$. Again, $|x - y| = |y - x|$. It is always true that $|a + b| \leqslant |a| + |b|$, so $|x - z| = = |(x - y) + (y - z)| \leqslant |x - y| + |y - z|$.

The most familiar distance function is the Euclidian metric $d : \mathbf{R}^2 \times \mathbf{R}^2 \longrightarrow \mathbf{R}_+$ defined by

$$d((x_1, x_2), (y_1, y_2)) = \sqrt{(x_1 - y_1)^2 + (x_2 - y_2)^2}$$

Here the metric axioms are easily verified by visual inspection. More generally, n-dimensional space $\mathbf{R}^n$ may be given the Euclidian metric

$$d((x_1, ..., x_n), (y_1, ..., y_n)) = \sqrt{(x_1 - y_1)^2 + ... + (x_n - y_n)^2}$$

A formal mathematical proof of the axioms for this metric is outlined in exercise 6.

Let us check the axioms for the 'city metric' $d : \mathbf{R} \times \mathbf{R} \longrightarrow \mathbf{R}_+$ given by

$$d((x_1, x_2), (y_1, y_2)) = |x_1 - y_1| + |x_2 - y_2|$$

Clearly $d((x_1, x_2), (y_1, y_2)) = 0$ if and only if $x_1 = y_1$ and $x_2 = y_2$ that is, if and only if $(x_1, y_1) = (x_2, y_2)$. The second axiom is immediate. For the triangle inequality (draw some pictures!)

$$\begin{aligned} & d((x_1, x_2), (y_1, y_2)) + d(y_1, y_2), (z_1, z_2)) \\ &= |x_1 - y_1| + |x_2 - y_2| + |y_1 - z_1| + |y_2 - z_2| \\ &= (|x_1 - y_1| + |y_1 - z_1|) + (|x_2 - y_2| + |y_2 - z_2|) \\ &\geqslant |x_1 - z_1| + |x_2 - z_2| \\ &= d((x_1, x_2), (z_1, z_2)). \end{aligned}$$

2 *Some mappings between metric spaces.* Let $(X, d), (Y, e)$ be metric spaces. A function $f : X \longrightarrow Y$ is an **isometry** *from* (X, d) *into* (Y, e) if for all $x_1, x_2 \in X$, $e(fx_1, fx_2) = d(x_1, x_2)$, that is, if f preserves distances. An isometry is automatically one-to-one, for if $x_1 \neq x_2$ then $e(fx_1, fx_2) = d(x_1, x_2) \neq 0$, which implies $fx_1 \neq fx_2$. f is an **isomorphism** $(X, d) \longrightarrow (Y, e)$ if f is an isometry and f is onto. In this case f^{-1} is also an isomorphism, since $d(f^{-1}y_1, f^{-1}y_2) = e(ff^{-1}y_1, ff^{-1}y_2) = e(y_1, y_2)$. It is intuitively clear that our notion of isomorphism is correct.

$f : (X, d) \longrightarrow (Y, e)$ is a *Lipschitz map* if there exists $\lambda > 0$ such that $e(fx_1, fx_2) \leqslant \lambda d(x_1, x_2)$ for all $x_1, x_2 \in X$. Lipschitz maps are much more general than isometries. Before giving examples, let us point out a simple principle that produces lots of metric spaces.

3 *Metric Subspaces.* Let (X, d) be a metric space and let A be any subset of X. Let $d_A : A \times A \longrightarrow \mathbf{R}_+$ be the restriction of d, that is $d_A(a_1, a_2) = d(a_1, a_2)$. Then (A, d_A) is a metric space as is trivial to verify, and we may call d_A the **subset metric**. It is clear, in fact, that d_A is the unique metric on A rendering the inclusion function $i : (A, d_A) \longrightarrow (X, d)$ an isometry.

Let us now give some examples of Lipschitz maps.

Consider $f: \mathbf{R} \longrightarrow \mathbf{R}$ defined by $fx = x^2$. Then $|fx - fy| = |x^2 - y^2| = |(x+y)(x-y)| = |x+y||x-y|$. As $x+y$ can be made arbitrarily large, $f: (\mathbf{R}, d) \longrightarrow (\mathbf{R}, d)$ is not a Lipschitz map if $d(x, y) = |x - y|$. On the other hand, let $A = \{x \in \mathbf{R} \mid |x| \leqslant 5\}$. Then $|x+y| \leqslant 10$ for $x, y \in A$ so that $f: (A, d_A) \longrightarrow (\mathbf{R}, d)$ is a Lipschitz map since $|fx - fy| \leqslant 10|x-y|$. Of course, every isometry is a Lipschitz map.

If $f: (X, d) \longrightarrow (Y, e)$ is constant ($fx_1 = fx_2$ for all $x_1, x_2 \in X$) then $e(fx_1, fx_2) = 0 \leqslant 0d(x_1, x_2)$ and f is Lipschitz, but f is not an isometry if X has at least two elements.

If $f: (X, d) \longrightarrow (Y, d_1)$ and $g : (Y, d_1) \longrightarrow (Z, d_2)$ are Lipschitz maps with $d_1(fx_1, fx_2) \leqslant \lambda d(x_1, x_2)$, $d_2(gy_1, gy_2) \leqslant \mu d_1(y_1, y_2)$ then $d_2(gfx_1, gfx_2) \leqslant \mu d_1(fx_1, fx_2) \leqslant \mu\lambda d(x_1, x_2)$ and gf is again Lipschitz. In this way, metric spaces and Lipschitz maps form a category. There is one grave disadvantage. Let A, B be the subsets $\{x \mid 0 \leqslant x \leqslant 5\}$, $\{x \mid 0 \leqslant x \leqslant 25\}$ of $\mathbf{R}$ with the (restricted) metrics d_A, d_B of $d(x, y) = |x - y|$ and define $f: (A, d_A) \longrightarrow (B, d_B)$ and $g : (B, d_B) \longrightarrow (A, d_A)$ by $fa = a^2$, $g(b) = \sqrt{b}$. Then f and g are Lipschitz maps satisfying $g \cdot f = \mathrm{id}_A$, $f \cdot g = \mathrm{id}_B$, so f is an isomorphism in the category of metric spaces and Lipschitz maps without being a 'true' isomorphism as defined earlier. The theory of sets with structure presented in Chapter 6 will provide a reason for preferring not to consider two metric spaces 'abstractly the same' if one contains two points 25 units apart and the other does not. Our philosophy, then, is that while Lipschitz maps have many uses they are not to be regarded, in general, as "homomorphisms of metric spaces". The preferred homomorphisms are as follows:

4 *The Category* **Met**. Let (X, d), (Y, e) be metric spaces. A **contraction** *from* (X, d) *to* (Y, e) is a function $f: X \longrightarrow Y$ satisfying $e(fx_1, fx_2) \leqslant d(x_1, x_2)$. It is clear that a contraction is Lipschitz (take $\lambda = 1$). By adapting the proof for Lipschitz maps above it is clear that $g \cdot f$ is a contraction when f and g are and, of course, $\mathrm{id}_X : (X, d) \longrightarrow (X, d)$ is a contraction. We obtain the category **Met** of metric spaces and contractions.

5 **PROPOSITION**: The isomorphisms in **Met** coincide with the isomorphisms defined earlier in 2.

Proof: If $f: (X, d) \longrightarrow (Y, e)$ is any isometry into, f is a contraction. If f is an isometry onto Y then f^{-1} is also an isometry (as proved in 2) so that f is an isomorphism in **Met** if f is an isomorphism in the earlier sense. Conversely, let $f: (X, d) \longrightarrow (Y, e)$ be an isomorphism in **Met**. Then $e(fx_1, fx_2) \leqslant d(x_1, x_2)$ because f is a contraction and $d(x_1, x_2) = d(f^{-1}fx_1, f^{-1}fx_2) \leqslant e(fx_1, fx_2)$ because f^{-1} is a contraction, so f is an isometry onto Y. □

We mention other categories of metric spaces which, it will develop, behave quite differently from **Met** with respect to some categorical constructions.

6 *The categories* **Met1** *and* **Met***. Let **Met1** be the category whose objects are metric spaces (X, d) *of diameter* $\leqslant 1$ (that is, $d(x, y) \leqslant 1$ for all x, y) and whose morphisms are contractions. Let **Met*** be the category whose objects are *metric spaces with base point* (X, d, x) (meaning (X, d) is a metric space and x "the base point" is an arbitrary element of X) and whose morphisms $f: (X, d, x) \longrightarrow (Y, e, y)$ are contractions $f: (X, d) \longrightarrow (Y, e)$ such that $fx = y$. It is clear that **Met*** is a category since if $fx = y$ and $gy = z$ then $gfx = gy = z$.

Exercises

1 In a metric space, a *circle of radius r and center* x_0 is the 'locus' (set) of points $\{x \in X \mid d(x, x_0) = r\}$. Let $d: \mathbf{R} \times \mathbf{R} \longrightarrow \mathbf{R}_+$ be the 'city metric' $|x_1 - y_1| + |x_2 - y_2|$ discussed in this section. Prove that every circle is a square.

2 Look up the role of Lipschitz maps in the "fundamental existence theorem" for differential equations.

3 A *metric-stretcher* is a function $f: \mathbf{R}_+ \longrightarrow \mathbf{R}_+$ satisfying $f(0) = 0$, f is one-to-one and $f(x) + f(y) \geqslant f(z)$ if $x + y \geqslant z$.
(a) If $d: X \times X \longrightarrow \mathbf{R}_+$ is a metric and $f: \mathbf{R}_+ \longrightarrow \mathbf{R}_+$ is a metric-stretcher then $d_f: X \times X \longrightarrow \mathbf{R}_+$ defined by $d_f(x, y) = f(d(x, y))$ is again a metric.
(b) Prove that $f(x) = x/(x + 1)$ is a metric-stretcher.
(c) If $a: \mathbf{R}_+ \longrightarrow \mathbf{R}_+$ is continuous and monotone decreasing (i.e., if $t \leqslant u$ then $a(u) \leqslant a(t)$) then

$$f(x) = \int_0^x a(t)dt$$

is a metric-stretcher.

4 Let (X, d) be a metric space. The **geometry of** (X, d) is the set G of all isomorphisms $f: (X, d) \longrightarrow (X, d)$ in **Met**. A **geometric figure** *in* (X, d) is a subset of X. The geometric figures A, B are **congruent** if there exists $g \in G$ with $g(A) = B$ (recall, $g(A) = \{g(a) \mid a \in A\}$), that is "$A$ can be moved to B via the 'rigid motion' g". (X, d) is **geometric** if all points are congruent, that is, for all x, y there exists $g \in G$ with $g(x) = y$.

(a) Prove that G is a group with id_X as unit and ordinary composition as multiplication. Use this to show that 'congruence' is an equivalence relation on geometric figures.

(b) If (X, d) is geometric, prove that any two circles (defined in exercise 1) with the same radius are congruent.

(c) Let $f : \mathbf{R}_+ \longrightarrow \mathbf{R}_+$ be a metric-stretcher (as defined in the previous exercise). Prove that (X, d) and (X, d_f) have the same geometry G.

(d) Let $X = \{a, b, c\}$ and define $d : X \times X \longrightarrow \mathbf{R}_+$ by the table

	a	b	c
a	0	1	3
b	1	0	2
c	3	2	0

Prove that (X, d) is a metric space, but is not geometric.

5 Let (X, d) be a metric space. Define a category $\mathbf{K}$ as follows. An object is an element x of X. A morphism $x \longrightarrow y$ is $\lambda \in \mathbf{R}_+$ such that $\lambda \geqslant d(x, y)$. If $\lambda : x \longrightarrow y$, $\mu : y \longrightarrow z$, $\mu \cdot \lambda : x \longrightarrow z$ is defined to be $\mu + \lambda$. Show that $\mathbf{K}$ is a category.

6 Prove that the Euclidian metric on $\mathbf{R}^n$

$$d(x, y) = \left[\sum_{j=1}^{n} (x_j - y_j)^2 \right]^{\frac{1}{2}}$$

satisfies the metric axioms. [Hint: Note that $d(x + \lambda y, 0) \geqslant 0$ to prove (recall your theory of quadratic equations) that

$$\left| \sum_{j=1}^{n} x_j y_j \right| \leqslant d(x, 0) d(y, 0)$$

Deduce that $d(x + y, 0) \leqslant d(x, 0) + d(y, 0)$.]

4.2 CONSTRUCTIONS IN CATEGORIES OF METRIC SPACES

1 *Products of Metric Spaces.* The one-element set 1 can have only one metric and only one base point and provides **Met**, **Met1** and **Met*** with a terminal object. Now consider a family $((X_i, d_i) \;\; i \in I)$ of metric spaces with $I \neq \emptyset$. Suppose that $p_i : (X, d) \longrightarrow (X_i, d_i)$ is a product in **Met**. Identifying an element of a set with the corresponding function from 1 and observing that any function $1 \longrightarrow (Y, e)$ is a contraction, the universal property induces a bijective correspondence

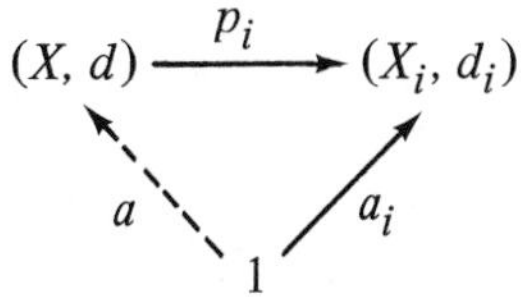

between elements a of X and I-tuples (a_i) with each $a_i \in X_i$. If (b_i) is another such I-tuple with corresponding $b \in X$ then, as each p_i is a contraction, we have $d_i(a_i, b_i) = d_i(p_i a, p_i b) \leqslant d(a, b)$ for all i. This proves that $\Pi(X_i, d_i)$ does not always exist in **Met**. For example, let $I = \mathbf{N}$ and let $(X_n, d_n) = (\mathbf{R}, d)$ for all n with $d(x, y) = x - y$. Then set $a_n = -n$, $b_n = n$. "$d(a_n, b_n) \leqslant d(a, b)$ for all n" reduces to the impossible condition "$d(a, b) \geqslant 2n$ for all n".

The difficulty we have raised is the only one:

2 A necessary and sufficient condition for the existence of $\Pi(X_i, d_i)$ in **Met** is that for every two I-tuples (a_i), (b_i) in the usual product set ΠX_i the I-tuple $d_i(a_i, b_i)$ of numbers has an upper bound N, (i.e. $d_i(a_i, b_i) < N$ for all i; such N may vary as the choice of (a_i) and (b_i) varies).

Proof: One direction in 2 has already been proved – we have shown that if I-tuples $d_i(a_i, b_i)$ exist without upper bound then the product cannot exist. Conversely, define $X = \Pi X_i$ to be the usual cartesian product set, and define

$$d((a_i), (b_i)) = \operatorname{Sup}\{d_i(a_i, b_i) \mid i \in I\}.$$

Let us verify that d is indeed a metric:

$$0 \leqslant d_i(a_i, b_i) \leqslant d(a, b) \quad \text{for all } i \qquad (\text{we write } a = (a_i),\ b = (b_i))$$

$$\begin{aligned} d(a, b) = 0 &\iff d_i(a_i, b_i) = 0 \quad \text{for all } i \\ &\iff a_i = b_i \qquad\quad\ \text{for all } i \\ &\iff a = b. \end{aligned}$$

That $d(a, b) = d(b, a)$ is trivial. Let $c = (c_i)$. Then for any fixed i we have

$$\begin{aligned} d(a, b) + d(b, c) &\geqslant d_i(a_i, b_i) + d_i(b_i, c_i) \\ &\geqslant d_i(a_i, c_i) \end{aligned}$$

Therefore $d(a, b) + d(b, c) \geqslant \operatorname{Sup}(d_i(a_i, c_i)) = d(a, c)$.

This proves that d is metric on X.

The usual coordinate projections $p_i : (X, d) \longrightarrow (X_i, d_i)$ are contractions since $d_i(a_i, b_i) \leqslant d(a, b)$. To prove the universal property, consider a family $h_i : (Y, e) \longrightarrow (X_i, d_i)$ of contractions.

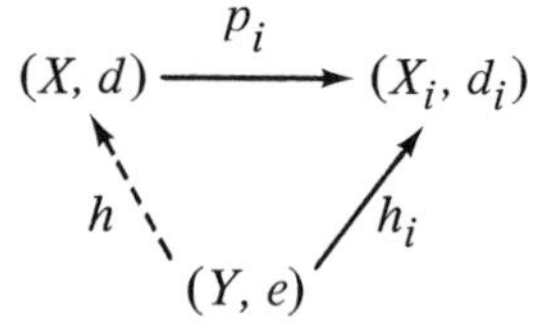

As expected, define $hy = (h_i y)$ so that h is the unique function with $p_i h = h_i$. We must show h is a contraction. For $y_1, y_2 \in Y$ we have $d(hy_1, hy_2) = \mathrm{Sup}(d_i(h_i y_1, h_i y_2))$. But as $d_i(h_i y_1, h_i y_2) \leqslant e(y_1, y_2)$ for all i, the upper bound $e(y_1, y_2)$ is at least as large as the least upper bound $d(hy_1, hy_2)$ and indeed h is a contraction. □

2 has two immediate corollaries. Any finite family has a product in **Met**. Also, **Met1** has products (since $\mathrm{Sup}(d_i(a_i, b_i)) \leqslant 1$ if each $d_i(a_i, b_i) \leqslant 1$).

By a construction similar to 2 we can see that, in contrast to **Met**, *every* family in **Met*** has a product, although it need not be built on the cartesian product set! Given (X_i, d_i, x_i) let X be the subset of all I-tuples (a_i) with each $a_i \in X$ such that the I-tuple of numbers $d_i(a_i, x_i)$ has an upper bound, and define $d((a_i), (b_i)) = \mathrm{Sup}\,\{d_i(a_i, b_i) \mid i \in I\}$. Then $x = (x_i)$ is in X.

Since there exist, by the definition of X, numbers s and t such that $d_i(a_i, x_i) \leqslant t$ for all i and $d_i(b_i, x_i) \leqslant s$ for all i then

$$d_i(a_i, b_i) \leqslant d_i(a_i, x_i) + d_i(x, b_i) = d_i(a_i, x_i) + d_i(b_i, x_i) \leqslant t + s$$

which proves that $d((a_i), (b_i))$ is a well-defined number. Let $p_i : (X, d, x) \longrightarrow (X_i, d_i, x_i)$ be the usual coordinate projections. We leave most of the verification remaining as an easy exercise in retracing the proof of 2, but mention one point in detail: to prove the universal property (see the diagram below)

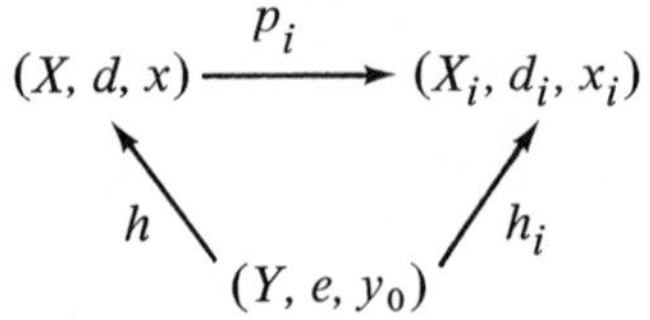

there is the new problem of proving that $hy \in X$ for all y. But $d_i(h_i y, x_i) = d_i(h_i y, h_i y_0) \leqslant e(y, y_0)$, so that $d_i(h_i y, x_i)$ has an upper bound, and $h(y) = (h_i y) \in X$.

Note that in general, X is a proper subset of the cartesian product set ΠX_i.

3 *Equalizers of pairs for metric spaces.* Let $f, g : (X, d) \longrightarrow (X', d')$ be a pair of morphisms in **Met**. Set $E = \{x \in X \mid fx = gx\}$ and give E the subset metric d_E as in 1.3. Let $i : (E, d_E) \longrightarrow (X, d)$ be inclusion. Then i is an isometry and thus a contraction. It is easy to check that $i = \mathrm{eq}(f, g)$ in **Met**. For let h be a contraction with $fh = gh$.

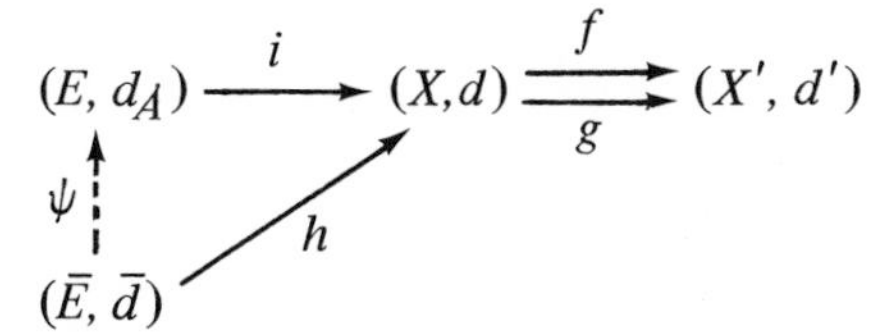

There exists a unique function ψ with $i\cdot\psi = h$, namely $\psi\bar{e} = h\bar{e}$. Also, $d_A(\psi\bar{e}_1, \psi\bar{e}_2) = d(h\bar{e}_1, h\bar{e}_2) \leqslant \bar{d}(\bar{e}_1, \bar{e}_2)$, so that ψ is a contraction.

If (X, d) is in **Met1** so is (E, d_A) so that, if also (X', d') is in **Met1**, $i = \mathrm{eq}(f, g)$ in **Met1**. If (X, d, x) and (X', d', x') are in **Met*** and $f(x) = x' = g(x)$ then (E, d_A, x) is in **Met*** and $i = \mathrm{eq}(f, g)$ in **Met***.

4 *Coproducts of metric spaces.* The empty set with its unique metric is the initial object in both **Met** and **Met1**. The empty set is not in **Met*** but 1 is clearly the initial object there. Binary coproducts do not exist in **Met**. In fact, $1 + 1$ does not exist. For suppose

$$1 \xrightarrow{\mathrm{in}_1} (X, d) \xleftarrow{\mathrm{in}_2} 1$$

were a coproduct diagram in **Met**. Let $Y = \{a, b\}$ and for each integer n let d_n be that metric $Y \times Y \longrightarrow \mathbf{R}_+$ such that $d_n(a, b) = n$. By the universal property, there exists a contraction ψ

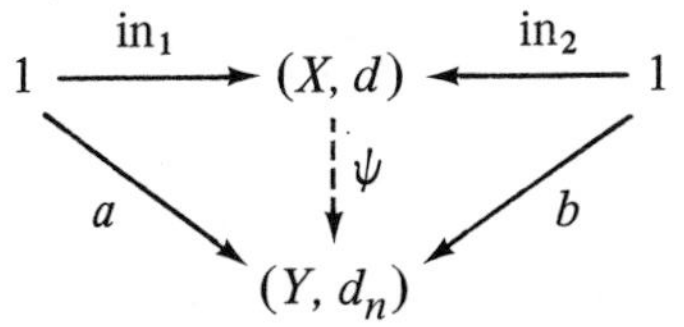

with $\psi(x_1) = a$ and $\psi(x_2) = b$ (where x_i is the element corresponding to $\mathrm{in}_i 1$ in X). Therefore, $d(x_1, x_2) \geqslant d_n(\psi x_1, \psi x_2) = n$ which is a contradiction since $d(x_1, x_2)$ is a fixed number and n can be chosen as large as we please.

These difficulties vanish in **Met1**. Given the family $\{(X_i, d_i) \mid i \in I\}$ in **Met1** with I non-empty, define X to be the disjoint union of the X_i, define $d : X \times X \longrightarrow \mathbf{R}_+$ by

$$d((x, i), (x', i')) = \begin{cases} d_i(x, x') & \text{if } i = i' \\ 1 & \text{if } i \neq i'. \end{cases}$$

We must check that d is a metric. The first two axioms are easy. The triangle inequality $d((x, i), (x', i')) + d((x', i'), (x'', i'')) \geqslant d((x, i), (x'', i''))$ breaks up naturally into four cases.

Case 1: $i = i' = i''$. Use the triangle inequality for d_i.

Case 2: $i = i' \neq i''$. $d_i(x, x') + 1 \geqslant 1$.

Case 3: $i = i'' \neq i'$. $d_i(x, x'') \leqslant 1 < 1 + 1$.

Case 4: i, i', i'' *distinct.* $1 + 1 > 1$.

The natural injections $\text{in}_i : (X_i, d_i) \longrightarrow (X, d)$, $\text{in}_i x = (x, i)$ are isometries and thus contractions. To check the universal property, let $h_i : (X_i, d_i) \longrightarrow (Y, e)$

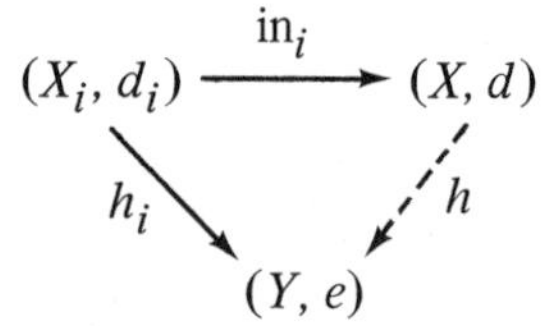

be a family of contractions and define h by $h(x, i) = h_i x$. Then h is the unique function such that $h \cdot \text{in}_i = h_i$. To prove h is a contraction, consider $(x, i), (x', i') \in X$. If $i = i'$, $e(h(x, i), h(x', i)) = e(h_i x, h_i x') \leqslant d_i(x, x') = d((x, i), (x', i'))$. If $i \neq i'$, $e(h(x, i), h(x', i')) \leqslant 1 = d((x, i), (x', i'))$. Thus (x, d) is truly the coproduct of the (X_i, d_i).

By a similar construction, **Met*** has all coproducts. Given the family (X_i, d_i, x_i) define $\tilde{X}_i$ to be the set $X_i - \{x_i\}$ of all elements of X_i except the basepoint x_i, let $\tilde{X}$ be the disjoint union of the $\tilde{X}_i$ and let $X = \tilde{X} \cup \{x\}$ where x denotes any new element not already in $\tilde{X}$. Define $d : X \times X \longrightarrow \mathbf{R}_+$ by

$$d(a, b) = \begin{cases} d_i(y, y') & \text{if } a = (y, i),\ b = (y', i) \\ d_i(y, x_i) + d_{i'}(x_{i'}, y') & \text{if } a = (y, i),\ b = (y', i'),\ i \neq i' \\ d_i(y, x_i) & \text{if } a = (y, i),\ b = x \text{ or } b = (y, i),\ a = x \\ 0 & \text{if } a = x = b. \end{cases}$$

We leave it as an exercise to the reader to complete the details.

5 *Coequalizers of pairs for metric spaces.* Let $f, g : (X, d) \longrightarrow (Y, e)$ be contractions. Let H be the class of all (Y', e', h) such that $h : (Y, e) \longrightarrow (Y', e')$ is a contraction with $hf = hg$. Define $R = \{(y_1, y_2) \mid hy_1 = hy_2 \text{ for all } h \in H\}$. R is an equivalence relation on Y. As usual, let $\eta_R : Y \longrightarrow Y/R$ denote canonical projection and define $d_R : Y/R \times Y/R \longrightarrow \mathbf{R}_+$ by

$$d_R(\alpha, \beta) = \text{Sup}\,\{e'(ha, hb) \mid (Y', e', h) \in H,\ [a] = \alpha,\ \ [b] = \beta\}.$$

Then d_R is a metric, $\eta_R : (Y, e) \longrightarrow (Y/R, d_R)$ is a contraction and $\eta_R = \text{coeq}(f, g)$ in **Met**. We leave the details, which are easy, to the reader. We note that the above construction is not very direct because we are unlikely to possess very specific information about H. This is an 'existence' proof.

Clearly, $(Y/R, d_R)$ is in **Met1** if (Y, e) is, so $\eta_R = \text{coeq}(f, g)$ in **Met1** if (X, d) and (Y, e) are in **Met1**.

Let $f, g : (X, d, x) \longrightarrow (Y, e, y)$ be morphisms in **Met***. As constructed above, let $\eta_R = \text{coeq}(f, g)$ in **Met**.

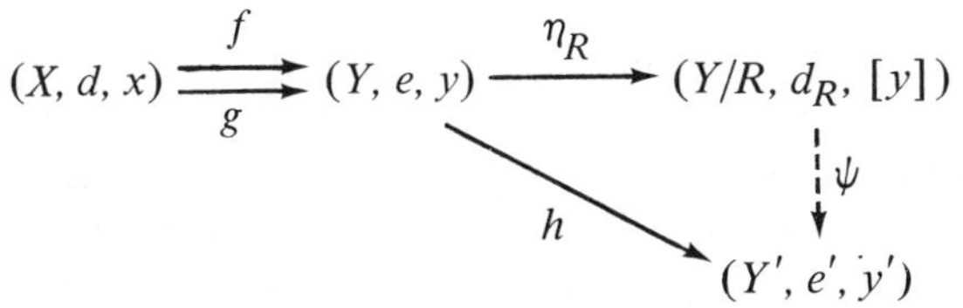

Define the base point of $(Y/R, d_R)$ to be $[y] = \eta_R(y)$. If $h \in$ **Met*** with $hf = hg$ then the unique $\psi \in$ **Met** with $\psi\eta_R = h$ satisfies $\psi[y] = y'$. Thus $\eta_R = \text{coeq}(f, g)$ in **Met***.

Exercises

1 For any fixed $M > 0$ let **Met***M* be the category of metric spaces of diameter $\leqslant M$. Show that **Met***M* has products, equalizers, coproducts and coequalizers.

2 An **extended** metric is a function $d : X \times X \longrightarrow \mathbf{R}_+ \cup \{\infty\}$ satisfying the same three laws as an ordinary metric (where, of course, $x \leqslant \infty$ and $x + \infty = \infty$ for all x in $\mathbf{R}_+ \cup \{\infty\}$. Let **Met∞** be the category of extended metric spaces and contractions. Show that **Met∞** has products, coproducts, equalizers and coequalizers.

3 Prove that, in **Met, Met1** and **Met***, a morphism is mono iff it is one-to-one whereas it is epi just in case it is onto.

4 Show that (onto, equalizer) and (coequalizer, one-to-one) are image factorization systems in **Met, Met1** and **Met***.

5 For any metric space (X, d) and $x \in X$ let $d_x : X \longrightarrow \mathbf{R}_+$ be the function $d_x(y) = d(x, y)$. Prove that $d_x : (X, d) \longrightarrow (\mathbf{R}_+, e)$ is a contraction where $e(a, b) = |a - b|$.

6 Consider $\mathbf{R}_+$ to be in **Met*** with the usual metric and 0 as base point. Show that every (X, d, x) in **Met*** is isomorphic to a subspace of a product (in **Met***) of copies of $\mathbf{R}_+$. [Hint: for any $y \in X$, $f_y : (X, d, x) \longrightarrow \mathbf{R}_+$, $z \mapsto d(y, z) - d(y, x)$ is a morphism in **Met***.]

7 Let (X_i, d_i) be a family in **Met1** such that each (X_i, d_i) is geometric in the sense of exercise 4 of Section 1. Prove that $\Pi(X_i, d_i)$ and $\amalg(X_i, d_i)$ are geometric.

4.3 TOPOLOGICAL SPACES

Not all interesting maps between metric spaces are contractions, or even Lipschitz. For example, we observed earlier that $x^2 : (\mathbf{R}, d) \longrightarrow (\mathbf{R}, d)$ is not a Lipschitz map. In calculus, one is led to deal with functions that are continuous (recall that f being a continuous function $\mathbf{R} \longrightarrow \mathbf{R}$ is an easy way to ensure that

$\int_a^b f(x)dx$ exists for every finite a, b). Paralleling the 'rigorous' definition of continuity in calculus texts, a funcıion $f: (X, d) \longrightarrow (Y, e)$ (where d and e are metrics) is *continuous at* $x_0 \in X$ if for all $\varepsilon > 0$ there exists $\delta > 0$ such that if $d(x, x_0) < \delta$ then $e(fx, fx_0) < \varepsilon$; i.e., "$f$ maps all points suitably close to x_0 to points near fx_0, thereby avoiding sudden jumps or discontinuities". f is *continuous* if f is continuous at all points x_0 in X.

Every Lipschitz function is continuous. To prove it, fix $x_0 \in X$, $\varepsilon > 0$. There exists $\lambda > 0$ such that $e(fx, fx_0) \leqslant \lambda d(x, x_0)$. Define $\delta = \varepsilon/\lambda$. Then if $d(x, x_0) < \delta$, $e(fx, fx_0) < \lambda \cdot (\varepsilon/\lambda) = \varepsilon$. However, note that $x^2 : (\mathbf{R}, d) \longrightarrow (\mathbf{R}, d)$, which we have observed is not Lipschitz, *is* continuous.

A key observation is that much less than metric structure is required to define 'continuous'. Given a metric space (X, d), x_0 in X and $\varepsilon > 0$, define the **sphere with center x_0 and radius** ε to be

$$S_\varepsilon(x) = \{x \text{ in } X \mid d(x, x_0) < \varepsilon\}.$$

Say that a subset $A \subset X$ is **open** *in* (X, d) if whenever x_0 is in A there exists $\varepsilon > 0$ such that $S_\varepsilon(x_0) \subset A$; i.e., "every point of A is in the interior of A".

If $f: (X, d) \longrightarrow (Y, e)$ is continuous, then $f^{-1}(B) = \{x \text{ in } X \mid fx \text{ is in } B\}$ is open in (X, d) whenever B is open in (Y, e). For let $x_0 \in f^{-1}(B)$. As $f(x_0)$ is in B there exists $\varepsilon > 0$ with $S_\varepsilon(fx_0) \subset B$. As f is continuous at x_0 there exists $\delta > 0$ such that "x in $S_\delta(x_0)$" implies "$f(x)$ in $S_\varepsilon(fx_0)$". But then $S_\delta(x_0) \subset f^{-1}(B)$ as desired.

The converse is also true. If $f^{-1}(B)$ is open whenever B is, then f is continuous. To prove it, fix $x_0 \in X$, $\varepsilon > 0$. The central observation is that $S_\varepsilon(fx_0)$ is open in (Y, e). For if $y \in S_\varepsilon(fx_0)$ then $d(fx_0, y) < \varepsilon$ and there exists a number $\gamma > 0$ with $d(fx_0, y) + \gamma < \varepsilon$. It follows from the triangle

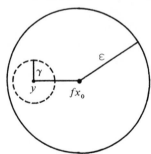

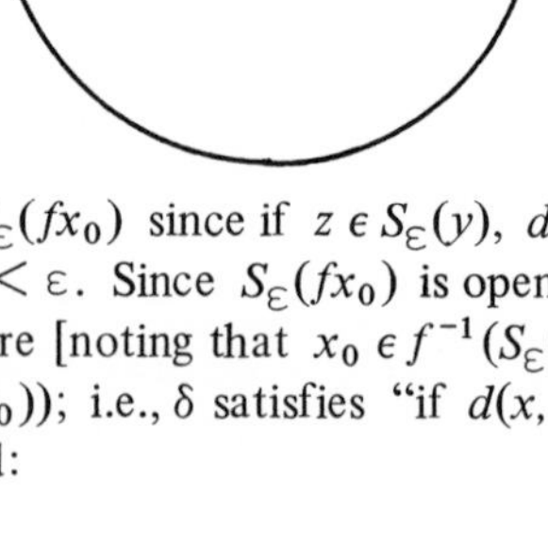

inequality that $S_\gamma(y) \subset S_\varepsilon(fx_0)$ since if $z \in S_\varepsilon(y)$, $d(fx_0, z) \leqslant d(fx_0, y) + d(y, z) < d(fx_0, y) + \gamma < \varepsilon$. Since $S_\varepsilon(fx_0)$ is open, $f^{-1}(S_\varepsilon(fx_0))$ is open by hypothesis and therefore [noting that $x_0 \in f^{-1}(S_\varepsilon(fx_0))$] there exists $\delta > 0$ with $S_\delta(x_0) \subset f^{-1}(S_\varepsilon(fx_0))$; i.e., δ satisfies "if $d(x, x_0) < \delta$ then $e(fx, fx_0) < \varepsilon$". So we have proved:

1 $f:(X,d)\longrightarrow(Y,e)$ is a continuous map of metric spaces if and only if $f^{-1}(B)$ is open in (X,d) whenever B is open in (Y,e). □

The point of 1 is that if our interest is only in continuity we can throw away the metrics if we keep the open sets.

If (X,d) is a metric space, the open sets in (X,d) satisfy certain properties. Clearly X is open, and the empty set is open (if a set A is not open one must exhibit an element $x_0\in A$ such that no $S_\varepsilon(x_0)\subset A$).

If A_1 and A_2 are open then $A_1\cap A_2$ is open: if $x_0\in A_1\cap A_2$ and $S_{\varepsilon_1}(x_0)\subset A_1$, $S_{\varepsilon_2}(x_0)\subset A_2$ then $S_\varepsilon(x_0)\subset A_1\cap A_2$ where ε is the minimum of ε_1 and ε_2.

Any union (not the disjoint union this time!) of open sets is open. For if $\{A_i \mid i\in I\}$ is a family of open sets and if

$$A=\cup A_i=\{x\in A \mid x\in A_i \text{ for at least one } i\in I\}$$

then if x_0 is in A it is in some A_i and, as A_i is open, there exists $\varepsilon>0$ with $S_\varepsilon(x_0)\subset A_i\subset A$.

If we call the collection of open sets of a metric space a **topology**, we are led to the following definition, which may hold whether or not X is provided with a metric:

2 **DEFINITION**: A **topology** τ on a set X is a collection of subsets of X which satisfies

(i) $\emptyset\in\tau$, $X\in\tau$

(ii) $A_1\cap A_2\in\tau$ if $A_1,A_2\in\tau$

(iii) If $\{A_i \mid i\in I\}$ is a family with each $A_i\in\tau$ then $\cup A_i\in\tau$.

A **topological space** is a pair (X,τ) where τ is a topology in the set X.

A **continuous mapping** $f:(X,\tau)\longrightarrow(Y,\sigma)$ *from* the topological space (X,τ) *to* the topological space (Y,σ) is a function $f:X\longrightarrow Y$ such that $f^{-1}(B)\in\tau$ whenever $B\in\sigma$.

$\mathrm{id}_X:(X,\tau)\longrightarrow(X,\tau)$ is obviously continuous. If $f:(X,\tau)\longrightarrow(Y,\sigma)$ and $g:(Y,\sigma)\longrightarrow(Z,\rho)$ are continuous then $gf:(X,\tau)\longrightarrow(Z,\rho)$ is continuous since, if $C\in\rho$, $g^{-1}(C)\in\sigma$ and $f^{-1}(g^{-1}(C))\in\tau$. But it is trivial to check that $(gf)^{-1}(C)=f^{-1}(g^{-1}(C))$. Thus:

3 **DEFINITION**: The category **Top** has as objects the topological spaces and as morphisms the continuous maps.

Exercises

1 (For terminology see exercise 3 of Section 1.) If $f:\mathbf{R}_+\longrightarrow\mathbf{R}_+$ is a metric stretcher and if (X,d) is a metric space, prove that (X,d) and (X,d_f) have the same topology.

2 A (possibly empty) family of morphisms $f_i : (X, \tau) \longrightarrow (X_i, \tau_i)$ in **Top** is *optimal* if whenever $g : (Y, \sigma) \longrightarrow X$ (i.e., (Y, σ) is a topological space and $g : Y \longrightarrow X$ is a function) is such that $f_i g : (Y, \sigma) \longrightarrow (X_i, \tau_i)$ is continuous for all i then $g : (Y, \sigma) \longrightarrow (X, \tau)$ is continuous.

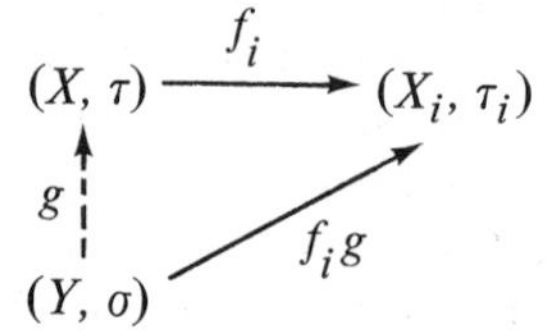

Given any set X and functions $f_i : X \longrightarrow (X_i, \tau_i)$ prove that the intersection of all the topologies containing $\cup \{f_i^{-1}(A) \mid A \in \tau_i\}$ is the unique topology τ on X such that $f_i : (X, \tau) \longrightarrow (X_i, \tau_i)$ is optimal. τ is called the *optimal lift of* $f_i : X \longrightarrow (X_i, \tau_i)$. The reader familiar with the 'subspace topology' on a subset A of a topological space (X, τ) should observe that this is just the optimal lift of the inclusion map of A. Also, the empty case is the 'discrete topology' on X.

3 Prove that every small diagram (see exercise 2.4.13) in **Top** has a limit. [Hint: Construct the limit in **Set** and use the unique optimal lift of the previous exercise.] The reader familiar with the 'Tychonoff product topology' should verify that it meshes with the construction of products as above.

4 A (possibly empty) family $f_i : (X_i, \tau_i) \longrightarrow (X, \tau)$ of morphisms in **Top** is *co-optimal* if whenever $g : X \longrightarrow (Y, \sigma)$ is such that $g f_i$ is continuous

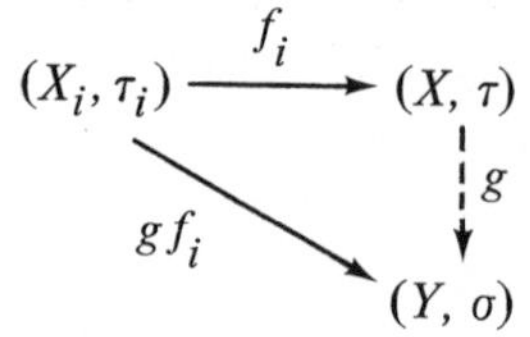

for all i then $g : (X, \tau) \longrightarrow (Y, \sigma)$ is continuous. Show that any $f_i : (X_i, \tau_i) \longrightarrow X$ has a unique co-optimal lift τ, which in the nonempty case is the set of all subsets A of X such that $f_i^{-1}(A) \in \tau_i$ for all i. The empty case is the 'indiscrete topology' on X.

5 Prove that every small diagram (see exercise 2.4.13) in **Top** has a colimit.

6 Prove that if $f : (X, \tau) \longrightarrow (Y, \sigma)$ is optimal (i.e. f is a one-element optimal family) then $\tau = \{f^{-1}(B) \mid B \in \sigma\}$. Prove "$f, g$ optimal $\Rightarrow g \cdot f$ optimal", and "$g \cdot f$ optimal $\Rightarrow f$ optimal".

7 Let $2 = \{0, 1\}$ and let σ be that topology on 2 in which $\{1\}$ is open and $\{0\}$ is not. $(2, \sigma)$ is called the **Sierpinski space**. For any (X, τ) in **Top** show

that the family of all continuous functions $(X, \tau) \longrightarrow (2, \sigma)$ is optimal. [Hint: For $A \subset X$, A is in τ if and only if its characteristic function $(X, \tau) \longrightarrow (2, \sigma)$ is continuous.]

Chapter 5

ADDITIVE CATEGORIES

We have said that the crucial concepts of category theory – *the categorical imperatives* – are arrows, structures and functors. In the first four chapters, we have shown how many concepts of mathematics can be placed in surprisingly general form when we express them in terms of *arrows*, rather than in terms of element-by-element definitions. In Part II, we shall introduce *functors*, and describe their many important roles, especially in the construction of free and cofree objects. But in the last two chapters of Part I we emphasize the structures which can be placed on collections of morphisms (an example of which is Chapter 5) and a general definition of 'structure' on sets (Chapter 6), steering a middle road between the general study of properties applicable in all categories, and the study of rather specific categories which occupied us in Chapters 3 and 4.

In the present chapter, we abstract from a number of general properties of **Vect**, the category of vector spaces and linear maps, to study *additive* (and *semi-additive*) categories. The most crucial property is that the product $A \times A$ is isomorphic to the coproduct $A + A$.

5.1 VECTORS AND MATRICES IN A CATEGORY

We start by taking a categorical view of the construction of matrices to represent linear maps in **Vect**. One of the intriguing properties of **Vect** (which we shall see to be shared by the semiadditive categories we study below) is that the coproduct $A_1 + \ldots + A_n$ of a finite collection of objects is, as an object, isomorphic to the product. Thus, $\mathbf{R}^n$ is a product in **Vect** via the projections $\pi_j : \mathbf{R}^n \longrightarrow \mathbf{R}$, $(x_1, \ldots, x_n) \mapsto x_j$ and is a coproduct in **Vect** via the injections $\mathrm{in}_i : \mathbf{R} \longrightarrow \mathbf{R}^n$, $x \mapsto (0, \ldots, x, 0, \ldots, 0)$ (where x occurs in the ith place); consult 2.4.3 and 2.4.4.

As a prelude to the situation in arbitrary categories, let us observe that a linear map $f : \mathbf{R}^3 \longrightarrow \mathbf{R}^2$ corresponds to a 2×3 *matrix* of linear maps $\mathbf{R} \longrightarrow \mathbf{R}$

$$(f_i^{\,j}) = \begin{pmatrix} f_1^{\,1} & f_2^{\,1} & f_3^{\,1} \\ f_1^{\,2} & f_2^{\,2} & f_3^{\,2} \end{pmatrix}$$

as follows. Because $\mathbf{R}^3 = \mathbf{R} + \mathbf{R} + \mathbf{R}$, f corresponds to

$$(f_1, f_2, f_3)$$

where $f_i = f \cdot \text{in}_i : \mathbf{R} \longrightarrow \mathbf{R}^2$. In turn, each f_i corresponds to

$$\begin{pmatrix} f_i^{\,1} \\ f_i^{\,2} \end{pmatrix}$$

where $f_i^{\,j} = \pi_j \cdot f_i : \mathbf{R} \longrightarrow \mathbf{R}$, because $\mathbf{R}^2 = \mathbf{R} \times \mathbf{R}$. Thus f corresponds to the 2×3 matrix $(f_i^{\,j})$.

While it is not necessary, for our purposes, to carry this example further, the reader may feel more comfortable if we do. Using the specific proofs of the universal properties in 2.4.3 and 2.4.4 we have

$$\begin{aligned} f(x, y, z) &= f_1(x) + f_2(y) + f_3(z) \\ &= \begin{pmatrix} f_1^{\,1}(x) + f_2^{\,1}(y) + f_3^{\,1}(z) \\ f_1^{\,2}(x) + f_2^{\,2}(y) + f_3^{\,2}(z) \end{pmatrix}. \end{aligned}$$

Now let us observe

1 FACT: If A is a vector space then the passage from $g : \mathbf{R} \longrightarrow A$ to $g(1)$ in A establishes a bijection between linear maps from $\mathbf{R}$ to A and elements in A, whose inverse assigns to $a \in A$ the linear map $\lambda \mapsto \lambda \cdot a$. □

Continuing our discussion of $f : \mathbf{R}^3 \longrightarrow \mathbf{R}^2$, each $f_i^{\,j} : \mathbf{R} \longrightarrow \mathbf{R}$ corresponds, as in 1, to a scalar $\lambda_i^{\,j}$ in $\mathbf{R}$. Thus f corresponds to the 2×3 matrix $(\lambda_i^{\,j})$ of scalars and

$$f(x, y, z) = \begin{pmatrix} \lambda_1^{\,1} \cdot x + \lambda_2^{\,1} \cdot y + \lambda_3^{\,1} \cdot z \\ \lambda_1^{\,2} \cdot x + \lambda_2^{\,2} \cdot y + \lambda_3^{\,2} \cdot z \end{pmatrix}.$$

which is the usual definition

$$\begin{pmatrix} \lambda_1^{\,1} & \lambda_2^{\,1} & \lambda_3^{\,1} \\ \lambda_1^{\,2} & \lambda_2^{\,2} & \lambda_3^{\,2} \end{pmatrix} \begin{pmatrix} x \\ y \\ z \end{pmatrix}$$

familiar from linear algebra.

Let us now see how far we can push the construction of vectors and matrices in an arbitrary category $\mathbf{K}$:

2 DEFINITION: Let A be an object of $\mathbf{K}$, and let $n \geqslant 1$ be an integer. An ***n*-ary codecomposition of** A is a coproduct diagram of the form $(\text{in}_i : A_i \longrightarrow A \mid i = 1, \ldots, n)$. The universal property establishes a bijective

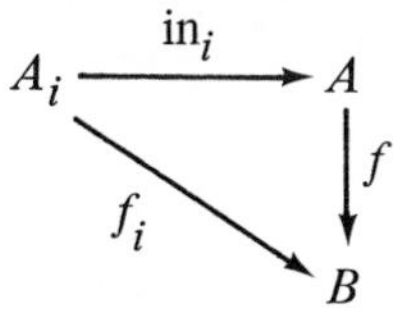

correspondence between morphisms $f : A \longrightarrow B$ and n-tuples

$(f_i : A_i \longrightarrow B \mid i = 1, ..., n)$, which we write as row 'vectors' $(f_1, ..., f_n)$.

3 **DEFINITION**: Dually to 2, an m-**ary decomposition of** A in **K** is a product diagram $\pi_j : A \longrightarrow B_j$, $j = 1, ..., m$. In the passage

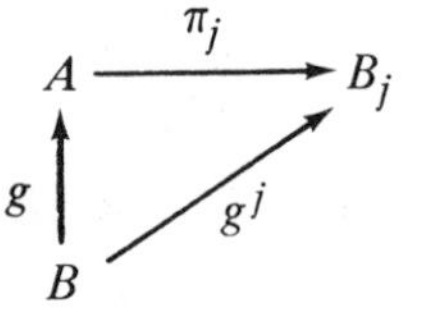

the n-vector $(g^j \mid 1 \leqslant j \leqslant m)$ corresponding to g will be written out with superscripts as a column vector

$$\begin{pmatrix} g^1 \\ \vdots \\ g^m \end{pmatrix}$$

More generally, consider an n-ary decomposition of A and an m-ary codecomposition of B. On the one hand, a morphism $f : A \longrightarrow B$ codecomposes into a row vector (f_i) and each f_i decomposes into a column vector $(f_i)^j$. We might also have first decomposed f and then codecomposed each f^j.

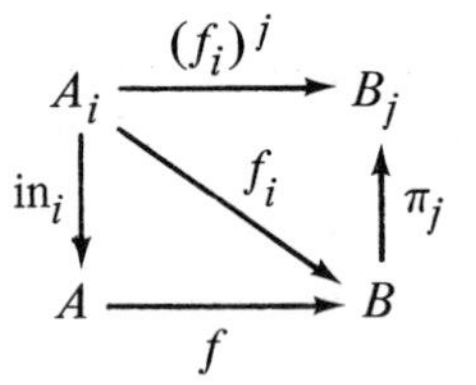

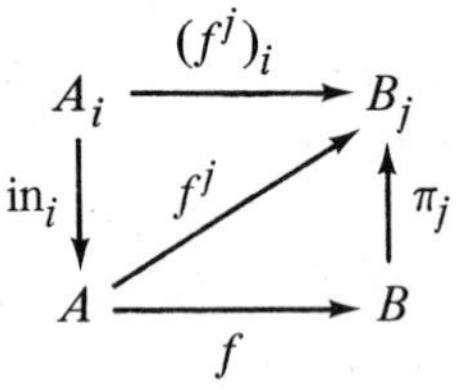

The following result makes it clear that $(f_i)^j = (f^j)_i$ for all i, j:

4 **PROPOSITION**: If $(\text{in}_i : A_i \longrightarrow A)$ is an n-ary codecomposition of A and if $(\pi_j : B \longrightarrow B_j)$ is an m-ary decomposition of B then for each family $f_i{}^j : A_i \longrightarrow B_j$, $1 \leqslant i \leqslant n$, $1 \leqslant j \leqslant m$, there exists a unique $f : A \longrightarrow B$ such that

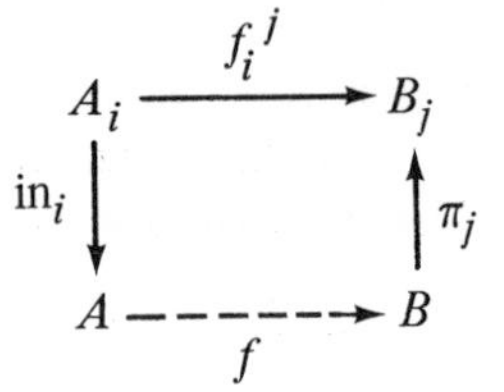

$f_i{}^j = \pi_j \cdot f \cdot \text{in}_i$ for all i, j.

Proof: Holding i fixed, there exists a unique f_i

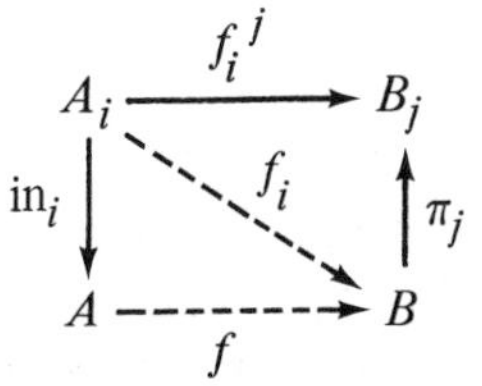

with $\pi_j \cdot f_i = f_i^{\,j}$ for all j. Now letting i vary, there exists a unique f with $f \cdot \text{in}_i = f_i$. Thus $\pi_j \cdot f \cdot \text{in}_i = \pi_j \cdot f_i = f_i^{\,j}$. If also $\pi_j \cdot g \cdot \text{in}_i = f_i^{\,j}$ then, fixing i and letting j vary, $g \cdot \text{in}_i = f_i$; but then $g = f$ as desired. □

Exercises

These two exercises presume the concept of a basis of a vector space.

1 Show that, in **Vect**, there is a bijective correspondence between bases of $\mathbf{R}^n$ and decompositions of the form $\pi_j : \mathbf{R}^n \longrightarrow \mathbf{R}$. [Hint: If $x_1, \ldots, x_n$ is a basis define $\pi_j(\Sigma \lambda_i x_i) = \lambda_j$.]

2 Show that, in **Vect**, there is a bijective correspondence between bases of $\mathbf{R}^n$ and codecompositions of the form $\text{in}_i : \mathbf{R} \longrightarrow \mathbf{R}^n$.

5.2 Abm-CATEGORIES

An **abelian monoid** is a triple $(X, +, 0)$ for which $+$ is an *abelian* (i.e. commutative) operation: $x + y = y + x$ for all x, y in X. Thus the function $+ : X \times X \longrightarrow X$ and the element 0 of X not only satisfy the usual conditions $x + (y + z) = (x + y) + z$ and $x + 0 = x = 0 + x$, but also $x + y = y + x$. Morphisms $f : (X, +, 0) \longrightarrow (X', +', 0')$ are required (as in **Mon**) to satisfy $f(x + y) = f(x) +' f(y)$ and $f(0) = 0'$. We denote the resulting category of abelian monoids by **Abm**.

The category **Vect** has (among others) the virtue that the set of morphisms between two objects forms an abelian monoid. For, given vector spaces X, Y if $f, g : X \longrightarrow Y$, then $f + g$ defined by $x \mapsto f(x) + g(x)$ is linear and $0 : X \longrightarrow Y$ with $x \mapsto 0$ for all $x \in X$ is also linear. Clearly **Vect**(X, Y) is an abelian monoid, since (for example) $(f + g)(x) = f(x) + g(x) = g(x) + f(x) = (g + f)(x)$. In other words, **Vect** admits an **Abm**-structure in the following sense:

1 **DEFINITION**: A category **K admits an Abm-structure** if there exists an assignment to each pair A, B of objects of **K** an abelian monoid structure $(+_{AB}, 0_{AB})$ on the set **K**(A, B) of **K**-morphisms from A to B subject to the

Distributive laws: Given $f : A \longrightarrow B$, $g_1, g_2 : B \longrightarrow C$, $h : C \longrightarrow D$,

$$(g_1 +_{BC} g_2) \cdot f = (g_1 \cdot f) +_{AC} (g_2 \cdot f)$$

$$h \cdot (g_1 +_{BC} g_2) = (h \cdot g_1) +_{BD} (h \cdot g_2).$$

Zero law: Given $f : A \longrightarrow B,\ g : C \longrightarrow D,$

$$0_{BC} \cdot f = 0_{AC}\,,$$
$$g \cdot 0_{BC} = 0_{BD}\,.$$

An **Abm**-category is a category with an **Abm**-structure.

In almost all cases we will write simply $f + g$, 0 instead of $f +_{AB} g$, 0_{AB}.

Of course, it is a special property to be an **Abm**-category. For example, **Set** is *not* an **Abm**-category (no 0_{AB}). For an even simpler example of a non-**Abm**-category, see exercise 3.

The more general concept of a **V**-category is introduced later in Chapter 9, where we shall extend our present analysis to the case in which **Abm** may be replaced by any category **V** of a suitable family.

2 **OBSERVATION**: The dual $\mathbf{K}^{\mathrm{op}}$ of an **Abm**-category **K** is an **Abm**-category. □

3 **DEFINITION**: A category **K** is **pointed** if there exists an assignment $0_{AB} \in \mathbf{K}(A, B)$ to each pair of objects subject to the *zero law* in 1 above.

Every **Abm**-category is pointed, but an assignment $(+_{AB}, 0_{AB})$ can satisfy the *distributive law* and fail the zero law with **K** pointed by $0'_{AB} \neq 0_{AB}$ (see exercise 3).

If 0_{AB}, $0'_{AB}$ are two assignments as in 3 then $0_{AB} = 0_{AB} \cdot 0'_{BB} = 0'_{AB}$, i.e. a category can be pointed in *at most* one way. Below we shall show that if **K** *has finite products and coproducts* it can admit at most one **Abm**-structure. In general, a category may have more than one **Abm**-structure (see exercise 12).

In introducing Section 1, we recalled that $A + A \cong A \times A$ for any vector space A. We devote the rest of this section to the following remarkable theorem:

4 **THEOREM**: Let **K** be a category in which $A + A$, $A \times A$ exist for every object A. Then the following two conditions are equivalent.

1. **K** admits an **Abm**-structure.
2. **K** is pointed (via 0_{AB}, say) and for all A, the map $\Delta : A + A \longrightarrow A \times A$ defined by $\pi_j \cdot \Delta \cdot \mathrm{in}_i = \delta_{ij}$ where
$$\delta_{ij} = \begin{cases} \mathrm{id}_A & \text{if } i = j \\ 0 & \text{if } i \neq j \end{cases}$$
is an isomorphism.

Moreover, if either of these conditions are true, the **Abm**-structure on **K** is unique.

Note that we have "$\Delta = \begin{pmatrix} 1 & 0 \\ 0 & 1 \end{pmatrix}$" in the matrix formulation of Section 1.

Proof: *1 implies 2.* Since **K** admits an **Abm**-structure, we may define Δ^{-1} by the rule

$$\Delta^{-1} : A \times A \longrightarrow A + A = \mathrm{in}_1 \cdot \pi_1 + \mathrm{in}_2 \cdot \pi_2 .$$

We must now check that Δ^{-1} is, as our notation suggests, really the inverse of Δ.

$$\begin{aligned} \Delta^{-1} \cdot \Delta \cdot \mathrm{in}_i &= (\mathrm{in}_1 \cdot \pi_1 + \mathrm{in}_2 \cdot \pi_2) \cdot \Delta \cdot \mathrm{in}_i \\ &= \mathrm{in}_1 \cdot \pi_1 \cdot \Delta \cdot \mathrm{in}_i + \mathrm{in}_2 \cdot \pi_2 \cdot \Delta \cdot \mathrm{in}_i \qquad \text{by the distributive law} \\ &= \mathrm{in}_1 \cdot \delta_{1i} + \mathrm{in}_2 \cdot \delta_{2i} = \mathrm{in}_i . \end{aligned}$$

Thus $\Delta^{-1} \cdot \Delta = \mathrm{id}_{A+A}$.

Similarly, $\pi_j \cdot \Delta \cdot \Delta^{-1} = \pi_j$ and $\Delta \cdot \Delta^{-1} = \mathrm{id}_{A \times A}$. Thus Δ is indeed an isomorphism $A \times A \cong A + A$ with inverse Δ^{-1}.

2 implies 1. $0 \in \mathbf{K}(A, B)$ is already defined since $\mathbf{K}$ is pointed. Addition in $\mathbf{K}(A, B)$ will be defined in two ways which will be seen to coincide. [Recall the row vector and column vector notation of Section 1.]

5 For $f, g : A \longrightarrow B$ define $f + g = (f, g) \cdot \Delta^{-1} \cdot \binom{\mathrm{id}}{\mathrm{id}}$

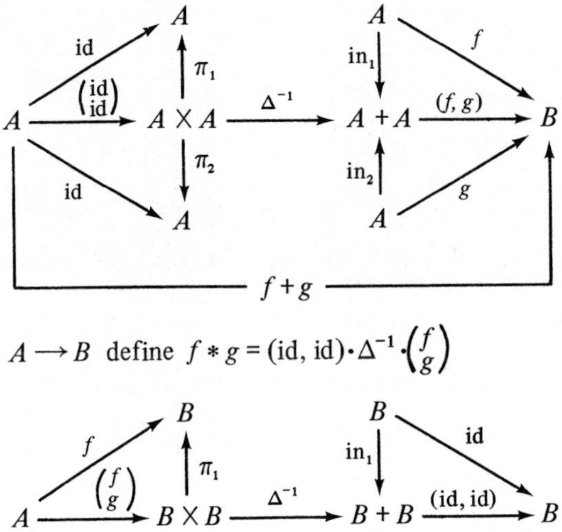

6 For $f, g : A \longrightarrow B$ define $f * g = (\mathrm{id}, \mathrm{id}) \cdot \Delta^{-1} \cdot \binom{f}{g}$

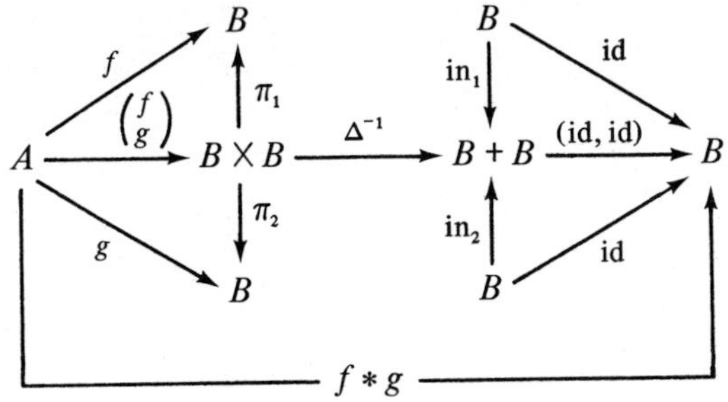

7 For $f : A \longrightarrow B$, $0 + f = 0 * f = f = f * 0 = f + 0$. The first two equalities will follow from the diagram below and the remaining two follow from a similar diagram.

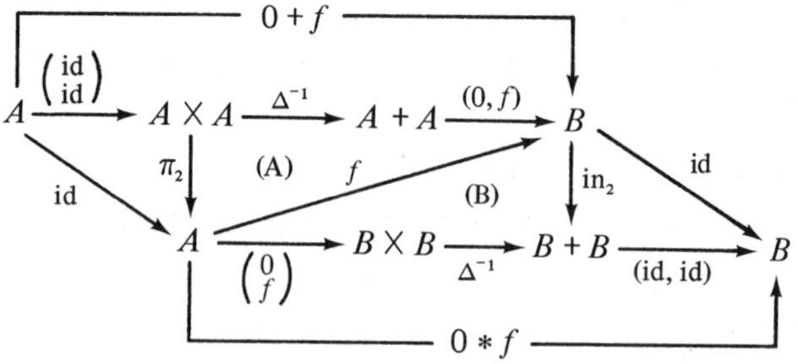

To prove that (A) is commutative, note that $f \cdot \pi_2 \cdot \Delta \cdot \text{in}_i = f \cdot \delta_{2i} = (0, f)\text{in}_i = (0, f)\Delta^{-1} \cdot \Delta \cdot \text{in}_i$ so that $f \cdot \pi_2 \cdot \Delta = (0, f)\Delta^{-1} \cdot \Delta$; (A) obtains by composing with Δ^{-1}. The proof that (B) commutes is similar. 7 is now established.

8 Given $f : A \longrightarrow B$, $g_1, g_2 : B \longrightarrow C$, $h : C \longrightarrow D$, then $h(g_1 + g_2) = hg_1 + hg_2$ and $(g_1 * g_2)f = g_1 f * g_2 f$.

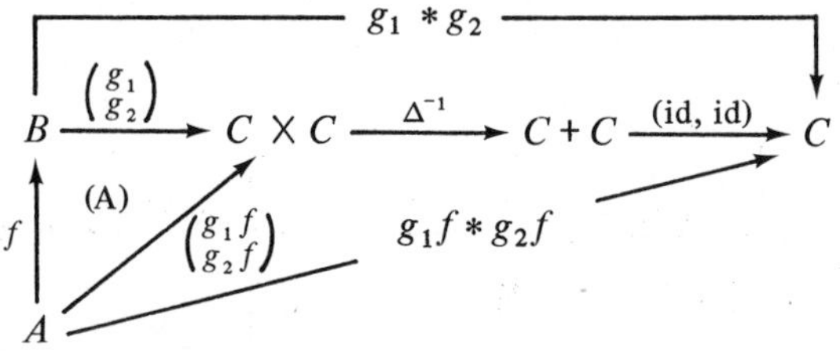

That (A) commutes is clear by composing with each $\pi_j : C \times C \longrightarrow C$ and so the second assertion is proved; the proof of the first is similar.

9 Given $w, x, y, z : A \longrightarrow B$, $(w * x) + (y * z) = (w + x) * (y + z)$. [This equality is known as the *middle four interchange*.]

$$\psi : A + A \longrightarrow B \times B = \begin{pmatrix} w & y \\ x & z \end{pmatrix},$$

i.e. ψ is determined uniquely (see 1.4) by

$$\pi_1 \cdot \psi \cdot \text{in}_1 = w, \qquad \pi_2 \cdot \psi \cdot \text{in}_1 = x,$$

$$\pi_1 \cdot \psi \cdot \text{in}_2 = y, \qquad \pi_2 \cdot \psi \cdot \text{in}_2 = z.$$

The crucial observation is that

$$\begin{pmatrix} (w, y) \\ (x, z) \end{pmatrix} = \begin{pmatrix} w & y \\ x & z \end{pmatrix} = \left(\begin{pmatrix} w \\ x \end{pmatrix} \begin{pmatrix} y \\ z \end{pmatrix} \right)$$

i.e., $\psi \cdot \text{in}_1 = \begin{pmatrix} w \\ x \end{pmatrix}$, $\psi \cdot \text{in}_2 = \begin{pmatrix} y \\ z \end{pmatrix}$ whereas $\pi_1 \cdot \psi = (w, y)$, $\pi_2 \cdot \psi = (x, z)$. It then follows from the diagrams

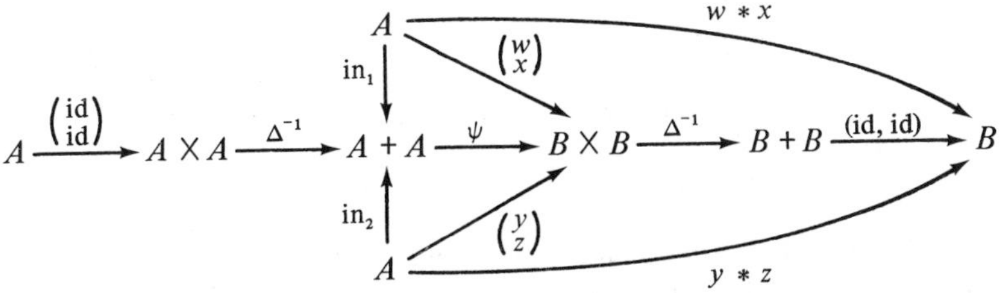

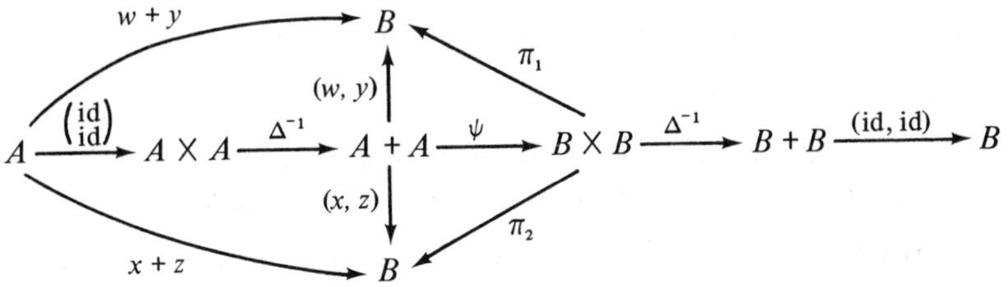

that $(w * x) + (y * z) = (\mathrm{id}, \mathrm{id}) \cdot \Delta^{-1} \cdot \psi \cdot \Delta^{-1} \cdot \binom{\mathrm{id}}{\mathrm{id}} = (w + y) * (x + z)$.

Armed with 6 – 9 we proceed as follows. In 9, set $x = 0 = y$. Then

$$w * z = (w + 0) * (0 + z) = (w * 0) + (0 * z) = w + z,$$

i.e. $+ = *$. 9 then becomes

$$(w + x) + (y + z) = (w + y) + (x + z).$$

Setting $w = 0 = z$, we get

$$x + y = y + x.$$

Setting $x = 0$ we derive

$$w + (y + z) = (w + y) + z.$$

Thus $(+, 0)$ is an abelian monoid structure on $\mathbf{K}(A, B)$. The distributive laws follow from 8 and the zero law is the pointedness of **K**. This completes the proof that conditions 1, 2 are equivalent.

It remains to show that if $(+', 0')$ is an **Abm**-structure on **K** then $+ = +'$ and $0 = 0'$. That $0 = 0'$ is clear from the discussion of 3. Since our definition of Δ does not depend on the choice of **Abm**-structure, the formula for its inverse in the proof that 1 implies 2 is valid for any **Abm**-structure. Thus, using the distributive laws for $+'$ we check that for any $f, g : A \longrightarrow B$

$$\begin{aligned} f+g &= \begin{pmatrix}\mathrm{id}\\ \mathrm{id}\end{pmatrix}\cdot\Delta^{-1}\cdot(f,g) \\ &= \left[\begin{pmatrix}\mathrm{id}\\ \mathrm{id}\end{pmatrix}\cdot\mathrm{in}_1\cdot\pi_1\cdot(f,g)\right] +' \left[\begin{pmatrix}\mathrm{id}\\ \mathrm{id}\end{pmatrix}\cdot\mathrm{in}_2\cdot\pi_2\cdot(f,g)\right] \\ &= f+' g. \end{aligned}$$ □

10 DEFINITION: A **semiadditive category** is a category with finite products and coproducts which admits a (unique) **Abm**-structure (+, 0). An **additive category** is a semiadditive category in which each abelian monoid $\mathbf{K}(A, B)$ is a group, i.e. for all $f: A \longrightarrow B$ there exists $-f: A \longrightarrow B$ with $f + (-f) = 0$.

Thus **Vect** is our standard example of an additive category.

Exercises

1 Let X, Y be monoids. Verify that, in general, given $f, g : X \longrightarrow Y \in$ **Mon**, $f{\cdot}g$ defined by $f{\cdot}g(x) = f(x){\cdot}g(x)$ is not necessarily in **Mon**.

2 Verify that **Vect** is additive and that **Abm** is semiadditive but not additive.

3 Let **K** be the category with one object A and a non-identity morphism u with $uu = u$. Composition also defines an abelian monoid structure $(+_{AA}, 0_{AA} = \mathrm{id}_A)$ on **K**. Verify that the distributive law holds and the zero law fails; and that **K** is pointed.

4 Verify that the distributive law implies the 0-law if $A + A$ and $A \times A$ exist. [Hint: Study the proof of 4.]

5 If **K** is an **Abm**-category then $A + B$ exists if and only if $A \times B$ exists. [Hint: If $A + B$ exists, define π_j on $A + B$ by $\mathrm{in}_i{\cdot}\pi_j = \delta_{ij}$.]

6 Construct the diagram to prove that $f = f * 0 = f + 0$ in 7. Also, prove that triangle (B) of 7 commutes.

7 Prove the first assertion of 8.

8 Prove that a semiadditive category is additive if and only if $\begin{pmatrix}\mathrm{id} & \mathrm{id}\\ 0 & \mathrm{id}\end{pmatrix} : A + A \longrightarrow A \times A$ is an isomorphism for all A.

9 Let $\mathbf{R}_+$ be the set of real numbers $x \geqslant 0$. A **semi vector space** is an abelian monoid $(X, +, 0)$ equipped with an action $\mathbf{R}_+ \times X \longrightarrow X$, $(\lambda, x) \mapsto \lambda{\cdot}x$ subject to the four laws of 2.1.1 II. Let **K** be the category of semi vector spaces whose morphisms are monoid homomorphisms satisfying $f(\lambda{\cdot}x) = x{\cdot}f(x)$.

(a) Verify that the positive quadrant of the plane is a semi vector space but not a vector space.

(b) Prove that **K** is semiadditive but not additive.

10 In any pointed category, the **kernel** of $f : A \longrightarrow B$ is the equalizer of f and 0_{AB}. Prove that in any additive category, the equalizer of f and g is the kernel of $f - g$.

11 A **commutative semiring** *with unit* is $(R, +, 0, \cdot, 1)$ where $(R, +, 0)$ is an abelian monoid, $(R, \cdot, 1)$ is a monoid, and the **distributive laws**

$$x \cdot (y + z) = (x \cdot y) + (x \cdot z)$$
$$(x + y) \cdot z = (x \cdot z) + (y \cdot z)$$

hold; if additionally $(R, +, 0)$ is a group, $(R, +, 0, \cdot, 1)$ is a **commutative ring with unit.** Show that the **R** and **Z** are rings whereas "reals $\geqslant 0$" and **N** are semirings (if addition and multiplication are defined as usual). Show that, if $(R, \cdot, 1)$ is considered as a category as in 3.1.3, each semiring is a semiadditive category and each ring is an additive category. "A semiring is a one-object semiadditive category."

12 Observe that the two-element set $R = \{0, a\}$ admits two distinct abelian monoid structures $+, +'$ with the same 0: $a + a = a$ and $a +' a = 0$. Observe that $(R, +, 0, \cdot, 0)$ and $(R, +', 0, \cdot, 0)$ are semirings where $x \cdot y = 0$. Conclude from exercise 11 that a category (which fails to have $A \times A$ and $A + A$) may have more than one **Abm**-structure.

13 Let **K** be the category with **N** as its only object and whose morphisms $\mathbf{N} \longrightarrow \mathbf{N}$ are arbitrary functions. Show that **N** admits a codecomposition of form $\text{in}_1, \text{in}_2 : \mathbf{N} \longrightarrow \mathbf{N}$ and a decomposition of form $\pi_1, \pi_2 : \mathbf{N} \longrightarrow \mathbf{N}$. Why is **K** not a semiadditive category?

14 Let A, B be sets. A **relation from** A **to** B, denoted $\alpha : A \rightharpoondown B$, is a subset α of $A \times B$. If also $\beta : B \rightharpoondown C$ define $\beta \cdot \alpha : A \rightharpoondown C$ by

$$\beta \cdot \alpha = \{(a, c) \mid \text{for some } b \in B, (a, b) \in \alpha \text{ and } (b, c) \in \beta\}.$$

(a) Show that **K**, whose objects are sets and whose morphisms are relations, is a category with composition as above.

(b) Given $\alpha : A \rightharpoondown B$ define $\alpha^* : B \rightharpoondown A$ by $\alpha^* = \{(b, a) \mid (a, b) \in \alpha\}$. Show that a diagram

$$A \xleftharpoondown{\pi_1} P \xrightharpoondown{\pi_2} B$$

is a product in R if and only if

$$A \xrightharpoondown{\pi_1^*} P \xleftharpoondown{\pi_2^*} B$$

is a coproduct in **K**.

(c) Show that $A \times B$ and $A + B$ exist in **K**. [Hint: Construct $A + B$ as a disjoint union and use (b).]

(d) Show that **K** is pointed via $0_{AB} = \emptyset$.

(e) Show that **K**(A, B) is an abelian monoid with unit $\emptyset$ via $\alpha \cup \beta$ and is even an abelian group via $(\alpha \cup \beta) - (\alpha \cup \beta)$.

(f) Discover why **K** is *not* semiadditive.

15 Let $(R, +, 0, \cdot, 1)$ be a commutative ring with unit (exercise 11). We say a set X equipped with two functions $X \times X \longrightarrow X : (x, x') \mapsto x + x'$ and $R \times X \longrightarrow X : (\lambda, x) \mapsto \lambda \cdot x$ is an R-**module** if it satisfies the conditions I and II of the definition of a vector space (2.1.1, with R replacing **R**). Then, mimicking the development of Section 2.1, we say a map $f : X \longrightarrow Y$ from one R-module to another is *linear* if $f(\lambda_1 x_1 + \lambda_2 x_2) = \lambda_1 f(x_1) + \lambda_2 f(x_2)$ for all x_1, x_2 in X and λ_1, λ_2 in R. Show that for any fixed commutative ring with unit, $\langle R$-modules and linear maps$\rangle$ forms a category, which we call R-**Mod**. Prove that R-**Mod** is additive.

Chapter 6

STRUCTURED SETS

In this chapter we use the language of category theory to provide a general definition of 'structure' on sets.

6.1 THE ADMISSIBLE MAPS APPROACH TO STRUCTURE

For motivation, consider the following break-down of the information defining the category **Poset**. For each set X, define **Poset**(X) to be the set of all partial orders $\leqslant$ on X. A poset is then a pair $(X, \leqslant)$ with $\leqslant$ in **Poset**(X). Notice that the concept of 'order-preserving map' $f : (X, \leqslant) \longrightarrow (X', \leqslant')$ is dependent on all four data $X, \leqslant, X', \leqslant'$. We might say that the system **Poset** determines – given X, X' with $\leqslant$ in **Poset**(X) and $\leqslant'$ in **Poset**(X') – a distinguished set of **admissible** functions from $(X, \leqslant)$ to $(X', \leqslant')$.

Again, we might define **Top**(X) to be all the topologies (Section 4.3) on the set X but our admissible map approach emphasizes the collection of *continuous* maps $(X, \tau) \longrightarrow (X', \tau')$. In the same spirit, **Vect**(X) would be all pairs of operation $+ : X \times X \longrightarrow X$ and $\mathbf{R} \times X \longrightarrow X$ which turn X into a vector space but our new approach emphasizes the collection of *linear* maps $(X, +, \cdot) \longrightarrow (X', +', \cdot')$. These examples can all be subsumed under the following general definition in which the system **C** associates a set **C**(X) of structures with each set X, and in which we make explicit the sets of admissible functions:

1 **DEFINITION**: A **category, C, of sets with structure** is given by the following two data and two axioms:

C assigns to each set X a set **C**(X) of **C-structures on** X. A **C-structure**, then, is a pair (X, s) with s in **C**(X). For each pair of sets (X, Y), **C** assigns a function

$$\mathbf{C}(X) \times \mathbf{C}(Y) \longrightarrow \mathbf{P}(Y^X) : s, t \mapsto \mathbf{C}(s, t)$$

where Y^X is, recall, the set of functions from X to Y and $\mathbf{P}(Y^X)$ is the set of its subsets. We write "$f : (X, s) \longrightarrow (Y, t)$" and say "$f$ is **admissible** in **C** from s to t" just in case f is in **C**(s, t). The axioms are

Admissible maps are composable: If $f : (X, s) \longrightarrow (Y, t)$ and $g : (Y, t) \longrightarrow (Z, u)$ then $g \cdot f : (X, s) \longrightarrow (Z, u)$.

Structure is abstract: If $f: X \longrightarrow Y$ is a bijection (i.e. an isomorphism of sets) and if t is in $\mathbf{C}(Y)$ there exists a unique s in $\mathbf{C}(X)$ with $f: (X, s) \longrightarrow (Y, t)$ and $f^{-1}: (Y, t) \longrightarrow (X, s)$.

The axioms are certainly true for **Poset**, for the composition of order-preserving maps is again order-preserving and if $f: X \longrightarrow X'$ is a bijection and $\leqslant'$ is in poset (X') then $\leqslant$ defined by $x_1 \leqslant x_2$ if and only if $f(x_1) \leqslant' f(x_2)$ is the unique partial order on X admitting f to $(X', \leqslant')$ and f^{-1} from $(X', \leqslant')$.

Note, however, that the category of metric spaces and Lipschitz maps is *not* a category of sets with structure because "structure is abstract" fails. Specifically, let (X, d) be the metric subspace $\{x \in \mathbf{R} : |x| \leqslant 5\}$ and let (Y, e) be the metric subspace $\{x \in \mathbf{R} : |x| \leqslant 25\}$. Then, as discussed on page 62, the map $f: (X, d) \longrightarrow (Y, e)$, $f(x) = x^2$ and its inverse $f^{-1}(y) = \sqrt{y}$ are Lipschitz, but d is *not* the unique metric with this property. Another such metric is $d'(x_1, x_2) = |x_1{}^2 - x_2{}^2|$. Here, d' is the unique metric making f an isometry.

2 **OBSERVATION**: A category $\mathbf{C}$ of sets with structure is in fact a category (also denoted $\mathbf{C}$) with *objects* all $\mathbf{C}$-structures (X, s) and with *morphisms* all $\mathbf{C}$-admissible maps $f: (X, s) \longrightarrow (Y, t)$. The composition is defined to be ordinary composition of functions. The identity $(X, s) \longrightarrow (X, s)$ is defined to be the identity function, id_X, of X.

Proof: Composition is well-defined by "admissible maps are composable". To prove that id_X is $\mathbf{C}$-admissible, observe first that by "structure is abstract" there exists (unique) s' in $\mathbf{C}(X)$ with $\mathrm{id}_X: (X, s') \longrightarrow (X, s)$ and $\mathrm{id}_X{}^{-1}: (X, s) \longrightarrow (X, s')$. By the axiom of composition applied to

$$(X, s) \xrightarrow{\mathrm{id}_X{}^{-1}} (X, s') \xrightarrow{\mathrm{id}_X} (X, s), \qquad \text{composite: } (X, s) \xrightarrow{\mathrm{id}_X} (X, s)$$

we have id_X is in $\mathbf{C}(s, s)$ as desired. □

Set, Vect, Mon, Grp, Met, Met1, Met* and **Top** are all conveniently viewed as categories of sets with structure. For example, define **Mon**(X) to be all pairs $(\circ, e)$ where $\circ: X \times X \longrightarrow X$ is an associative multiplication with $e \in X$ as identity. Let admissible maps be monoid homomorphisms. To prove "structure is abstract" take a bijection $f: X \longrightarrow Y$ and a structure $(*, u)$ in **Mon**(Y). We may then define the requisite $(\circ, e)$ in **Mon**(X) by $x \circ y = f^{-1}(f(x) * f(y))$ and $e = f^{-1}(u)$.

3 **DEFINITION**: An **isomorphism** of the categories $\mathbf{C}$, $\mathbf{D}$ of sets with structure is given by bijections $\psi_X: \mathbf{C}(X) \longrightarrow \mathbf{D}(X)$ with the property that for all s

in $\mathbf{C}(X)$, t in $\mathbf{C}(Y)$ and maps $f: X \longrightarrow Y$, we have that $f: (X, s) \longrightarrow (Y, t)$ is true iff $f: (X, \psi_X(s)) \longrightarrow (Y, \psi_Y(t))$ is true.

It is easy to give examples of such isomorphisms which support the feeling that the definition is a good one. For example, an alternate definition of 'group', as opposed to that of 3.1.1, is a quadruple $(X, \circ, e, i)$ where $(X, \circ, e)$ is a monoid and $i: X \longrightarrow X$ is a function satisfying $x \circ i(x) = e = i(x) \circ x$ for all x. A homomorphism $f: (X, \circ, e, i) \longrightarrow (X', \circ', e', i')$ must satisfy $f(x) \circ' f(y) = f(x \circ y)$, $f(e) = e'$, $f(ix) = i'(fx)$. Most textbooks on algebra would agree that this formulation is 'equivalent' to 3.1.1. A more precise definition is that the passage ψ_X from $(\circ, e)$'s on X as in 3.1.1 to $(\circ, e, i)$'s on X as above (defined by $i(x) = x^{-1}$) is an isomorphism of categories of sets with structure.

Sometimes categories of sets with structure that at first appear quite different are seen to be isomorphic. Here is such an example:

4 *Posets as topological spaces.* Let $\mathbf{C}$ be the category of sets with structure whose objects are those topological spaces (X, τ) satisfying, in addition to (i) – (iii) of 4.3.2 the conditions:

(iv) If $(A_i \mid i \in I)$ is a family with each $A_i \in \tau$ then $\cap A_i \in \tau$.

(v) (X, τ) is a **T0 space**, that is, if $x \neq y$ there exists $A \in \tau$ with $x \in A$, $y \notin A$ or with $y \in A$, $x \notin A$.†

Let the $\mathbf{C}$-admissible maps be the continuous maps. Then $\mathbf{C}$ is isomorphic, as a category of sets with structure, to **Poset**.

Proof: Define $\psi_X : \mathbf{Poset}(X) \longrightarrow \mathbf{C}(X)$ by

$$\psi_X(\leqslant) = \{A \subset X \mid \text{whenever } x \leqslant a \text{ and } a \in A \text{ then } x \in A\}$$
$$\psi_X^{-1}(\tau) = \{(x, y) \mid \text{whenever } y \in A \text{ and } A \in \tau \text{ then } x \in A\}.$$

It is then straightforward to check that ψ is an isomorphism of categories of sets with structure. □

Exercises

1 Verify in detail that **Vect**, **Grp**, and **Top** are categories of sets with structure.

2 Prove that the category of metric spaces and continuous maps is not a category of sets with structure.

3 A **preordered set** is a pair $(X, \leqslant)$ where $\leqslant$ is a binary relation which is reflexive and transitive (but not necessarily antisymmetric). Just as for posets, define $f: (X, \leqslant) \longrightarrow (X', \leqslant')$ to be admissible just in case $x \leqslant y$

---

† Note that $\mathbf{R}$ with the usual Euclidian topology induced by the metric $|x - y|$ certainly has this property, for given $x \neq y$, we may take $A = S_\epsilon(x)$ for any $\epsilon < |x - y|$. This is called a *separation* property in topology.

implies $f(x) \leqslant f(y)$. Show that the resulting category is a category of sets with structure which is isomorphic to the category of those topological spaces in which every intersection of open sets is open.

4 Observe that the definition of "categories of sets with structure" depends only upon the fact that **Set** is a category. Define "category of **K**-objects with structure" for an arbitrary category **K**. Prove that **Grp** is a category of monoids with structure and that **Vect** is a category of groups with structure. If **C** is a category of **K**-objects with structure suggest the natural definition of $\mathbf{C}^{\mathrm{op}}$ as a category of $\mathbf{K}^{\mathrm{op}}$-objects with structure.

6.2 OPTIMAL FAMILIES

In familiar categories of sets with structure, limits, colimits, substructures and quotient structures can often be achieved or defined by providing the corresponding construction in **Set** and adjoining the appropriate structure. In this section we study a general technique to carry out such a program.

1 **DEFINITION:** Let **C** be a category of sets with structure. An I-indexed family $f_i : (X, s) \longrightarrow (X_i, s_i)$ of admissible maps is **optimal** if whenever we have a diagram

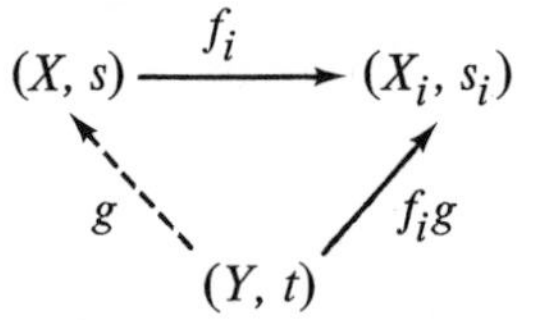

where (Y, t) is a **C**-structure and $g : Y \longrightarrow X$ is a function such that $f_i g : (Y, t) \longrightarrow (X_i, s_i)$ is admissible for all $i \in I$; then $g : (Y, t) \longrightarrow (X, s)$ is admissible.

Dually, $f_i : (X_i, s_i) \longrightarrow (X, s)$ is **co-optimal** if whenever (Y, t) and $g : X \longrightarrow Y$ are such that $g \cdot f_i : (X_i, s_i) \longrightarrow (Y, t)$ is admissible,

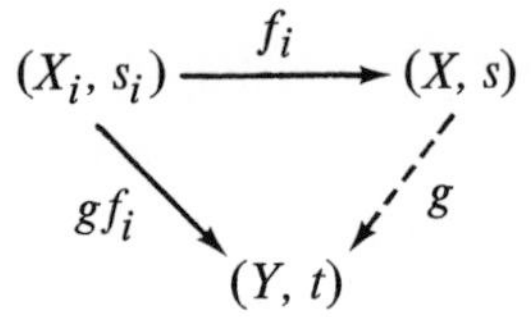

then $g : (X, s) \longrightarrow (Y, t)$ is admissible.

Given a family

$$X \xrightarrow{f_i} (X_i, s_i)$$

(i.e. (X_i, s_i) are **C**-structures and $f_i : X \longrightarrow X_i$ are functions), (f_i) **has an optimal lift** if there exists s in $\mathbf{C}(X)$ such that $f_i : (X, s) \longrightarrow (X_i, s_i)$ is optimal.

Dually,

$$(X_i, s_i) \xrightarrow{f_i} X$$

has a co-optimal lift if there exists s in $\mathbf{C}(X)$ such that $f_i : (X_i, s_i) \longrightarrow (X, s)$ is co-optimal.

2 **LEMMA**: Optimal and co-optimal lifts, when they exist, are unique.

Proof: Let $f_i : (X, s) \longrightarrow (X_i, s_i)$, $f_i : (X, s') \longrightarrow (X_i, s_i)$ be optimal families. As the first is admissible and the second is optimal, $\mathrm{id}_X : (X, s) \longrightarrow (X, s')$ is admissible. Similarly, $\mathrm{id}_X : (X, s') \longrightarrow (X, s)$ is admissible. Since $\mathrm{id}_X : (X, s) \longrightarrow (X, s)$ is admissible, it follows from the uniqueness assertion of "structure is abstract" that $s = s'$. The uniqueness of co-optimal lifts is dual. □

We now examine several examples.

3 **DEFINITION**: Let $\mathbf{C}$ be a category of sets with structure, let (X, s) be a $\mathbf{C}$-structure and let $A \subset X$. A is a **$\mathbf{C}$-substructure** if the inclusion map $\mathrm{inc} : A \longrightarrow X$, $a \mapsto a$ has an optimal lift t in $\mathbf{C}(A)$. It is also conventional to say "(A, t) is a substructure of (X, s)".

4 **EXAMPLE**: Let $\mathbf{C}$-structure mean vector space. If (X, s) is a vector space and $A \subset X$ then A is a subspace in the sense of 2.1.2 if and only if A is a $\mathbf{C}$-substructure. To prove this, we make the following observations:

(i) If $\mathrm{inc} : (A, t) \longrightarrow (X, s)$ is admissible for some t then for $a, a' \in A$,

$$a +_t a' = \mathrm{inc}(a +_t a') = \mathrm{inc}(a) +_s \mathrm{inc}(a') = a +_s a'$$

and, similarly, there is no ambiguity over $\lambda \cdot a$. Thus A must be a subspace in the sense of 2.1.2 which, as a vector space, has structure t.

(ii) If A is a subspace of (X, s) as in 2.1.2, then A is a vector space (A, t) and $\mathrm{inc} : (A, t) \longrightarrow (X, s)$ is optimal. Admissibility is clear.

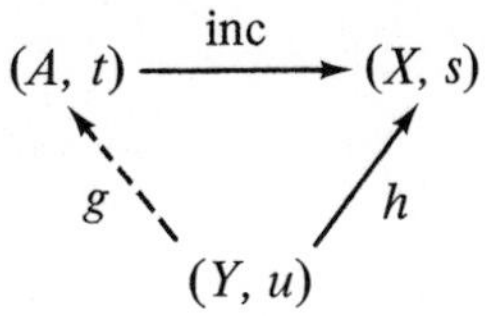

Now suppose $g : Y \longrightarrow A$, (Y, u) are such that $h = \mathrm{inc} \cdot g$ is admissible. Then

$$\begin{aligned} g(y +_u y') &= \mathrm{inc}(g(y +_u y')) \\ &= h(y) +_s h(y') \\ &= g(y) +_t g(y') \qquad \text{(reasoning as in (i))}. \end{aligned}$$

Similarly, one may show that $g(\lambda \cdot y) = \lambda \cdot g(y)$.

Definition 3 seems unnecessarily complicated in the context of 4 because of 4(i) which points out that inclusion maps of vector spaces have at most one structural lift. Compare, however, the metric space situation, where there exist non-optimal structural lifts:

5 **EXAMPLE**: Let **C**-structure mean metric space with admissible maps as in 4.1.4. Then every subset A of (X, d) is a substructure and the optimal structure $d' \in \mathbf{C}(A)$ is obtained by restricting d, i.e. $d'(a, b) = d(a, b)$.

It is optimality that forces d' to be the restriction of d for, in general, there are many other structures e in $\mathbf{C}(A)$ such that $\text{inc} : (A, e) \longrightarrow (X, d)$ is admissible. For example let $X = \{x \in \mathbf{R} \mid 0 \leqslant x \leqslant 1\}$, $d(x, y) = |x - y|$, $A = X$, $e(x, y) = 1$ if $x \neq y$, 0 if $x = y$. Then $\text{id} : (X, e) \longrightarrow (X, d)$ is admissible but e is not optimal since d is optimal and $d \neq e$.

We now investigate the relationship between optimality and constructions of products and equalizers.

6 **THEOREM**: Let **C** be a category of sets with structure and let (X_i, s_i) be a family of **C**-structures. Let $\pi_i : X \longrightarrow X_i$ be the usual product in **Set**. If

$$X \xrightarrow{\pi_i} (X_i, s_i)$$

has an optimal lift $s \in \mathbf{C}(X)$, then

$$(\pi_i : (X, s) \longrightarrow (X_i, s_i)) = \Pi(X_i, s_i)$$

the product in **C**.

Proof: Let $f_i : (Y, t) \longrightarrow (X_i, s_i)$ be admissible.

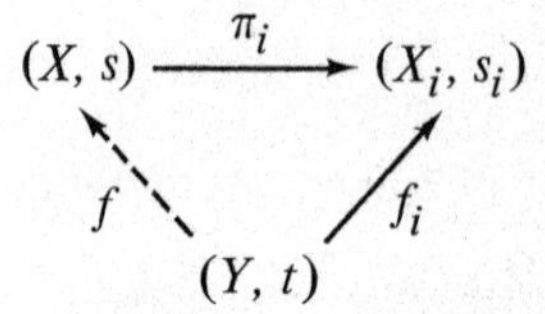

Then there exists a unique function $f : Y \longrightarrow X$ such that $\pi_i f = f_i$ for all i. As each f_i is admissible and s is optimal, f is admissible. □

Theorem 6 is very often applicable (see exercise 8). However, even though **Met*** has products they cannot always be constructed by 6 (see exercise 9).

7 **THEOREM**: Let **C** be a category of sets with structure, let $f, g : (X, s) \longrightarrow (Y, t)$ be admissible and let $i : E \longrightarrow X = \text{eq}(f, g)$ in **Set**. Then, if $i : E \longrightarrow (X, s)$ has an optimal lift u in $\mathbf{C}(E)$, $i : (E, u) \longrightarrow (X, s) = \text{eq}(f, g)$ in **C**.

Proof: Let $h : (T, v) \longrightarrow (X, s)$ be admissible such that $fh = gh$.

$$(E, u) \xrightarrow{\quad i \quad} (X, s) \underset{g}{\overset{f}{\rightrightarrows}} (Y, t)$$

$$(E, u) \overset{h'}{\nwarrow} \quad \nearrow h$$
$$(T, v)$$

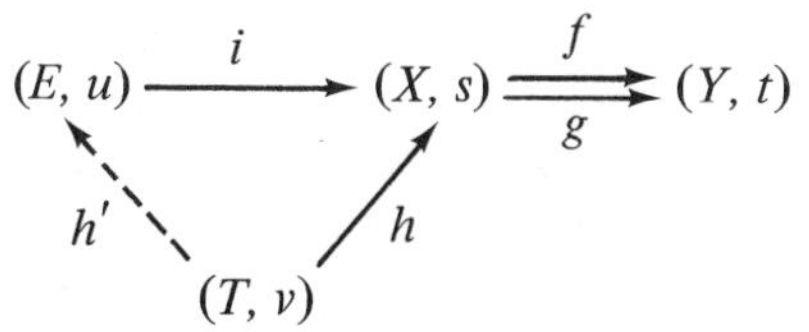

There exists a unique function h' with $ih' = h$. Since h is admissible and u is optimal, h' is admissible.   □

Most familiar categories have equalizers by the construction of 7 (see exercise 10). Notice that such equalizers are substructures.

The theory of 3, 6 and 7 dualizes in a natural way:

8   DEFINITION: Let (X, s) be a **C**-structure and let E be an equivalence relation on X. Then E is a **C-congruence** and X/E is a **quotient structure of** (X, s) if the canonical map

$$\eta_E : (X, s) \longrightarrow X/E$$

(as in 1.3.4) has a co-optimal lift t in $\mathbf{C}(X/E)$. It is usual to say that $(X/E, t)$ is a **quotient structure** of (X, s).

9   THEOREM: Let **C** be a category of sets with structure, let (X_i, s_i) be a family of **C**-structures and let $\mathrm{in}_i : X_i \longrightarrow X = \amalg X_i$ in **Set**. Then if

$$\mathrm{in}_i : (X_i, s_i) \longrightarrow X$$

has a co-optimal lift s in $\mathbf{C}(X), (\mathrm{in}_i : (X_i, s_i) \longrightarrow (X, s)) = \amalg(X_i, s_i)$, the coproduct in **C**.   □

10   THEOREM: Let **C** be a category of sets with structure, let $f, g : (X, s) \longrightarrow (Y, t)$ be admissible and let $h : Y \longrightarrow Z = \mathrm{coeq}(f, g)$ in **Set**. Then if

$$(Y, t) \xrightarrow{\quad h \quad} Z$$

has a co-optimal lift u in $\mathbf{C}(Z)$, $h : (Y, t) \longrightarrow (Z, u) = \mathrm{coeq}(f, g)$ in **C**.   □

Exercises

1   Verify that co-optimal lifts are unique.

2   Prove that s in $\mathbf{C}(X)$ is the optimal lift of the empty family of maps out of X if and only if for all (Y, t) and $f : X \longrightarrow Y$, $f : (X, s) \longrightarrow (Y, t)$ is admissible. Such an s is called the **discrete structure on** X. The dual notion is **indiscrete structure**. The terminology comes from topology; the discrete topology is $\{S \mid S \subset X\}$ and the indiscrete topology is $\{\emptyset, X\}$.

3 Convince yourself that one must use categories of $\mathbf{Set}^{op}$-objects with structure (as in exercise 4 of Section 1) to formalize the duality between optimality and co-optimality. [Hint: If **C** is a category of sets with structure, $\mathbf{C}^{op}$ is a category of $\mathbf{Set}^{op}$-objects with structure.]

4 Rework Example 4 with $\mathbf{C} = \mathbf{Mon}$ and $\mathbf{C} = \mathbf{Grp}$.

5 In the context of Example 5 verify that d' is optimal.

6 Let **C** be topological spaces and continuous maps, as in 4.3.3. If (X, τ) is a **C**-structure and $A \subset X$ show that $\mathbf{S} = \{T \cap A \mid T \in \tau\}$ in $\mathbf{C}(A)$ is optimal. **S** is called the **subspace topology** in the literature.

7 Verify that Theorem 6 yields a terminal object in **C** in the empty case.

8 Verify that products in **Vect, Grp, Top, Met1**, and finite products in **Met**, arise as in Theorem 6.

9 Consider $(\mathbf{R}, |x - y|, 0)$ in **Met***. Show that, if $X = \mathbf{R}^I$ with I infinite, there is no structure (d, x) in $\mathbf{Met^*}(X)$ such that each $\pi_i : (X, d, x) \longrightarrow (\mathbf{R}, |x - y|, 0)$ is admissible. Show that **Met*** has products.

10 Show that **Vect, Grp, Top, Met, Met*** and **Met1** have equalizers as in Theorem 7.

11 In **Mon, Grp** and **Vect** show that if E is an equivalence relation on (X, s) then E is a congruence if and only if E is a substructure of $(X, s) \times (X, s)$.

12 In **Top**, show that every equivalence relation E on (X, τ) is a congruence. [Hint: If $\eta : X \longrightarrow X/E$, define σ in $\mathbf{C}(X/E)$ by $\sigma = \{A \subset X/E \mid \eta^{-1}(A) \in \tau\}$. σ is called the **quotient topology** in the literature.]

13 Prove Theorems 9 and 10.

14 Prove that Theorem 9 does not, in general, construct coproducts in **Abm**, **Met** or **Met*** but does construct coproducts in **Met1**.

15 Prove that in **Top** every family $f_i : X \longrightarrow (X_i, \tau_i)$ has an optimal lift. [Hint: Consider the intersection of all topologies containing $f_i^{-1}(\tau_i)$ for every i.]

16 Prove that in **Top** every family $f_i : (X_i, \tau_i) \longrightarrow X$ has a co-optimal lift. [Hint: Consider $\{A \subset X \mid f_i^{-1}(A) \in \tau_i \text{ for all } i\}$.]

17 State and prove a generalization of Theorems 6 and 7 for limits of arbitrary diagrams in **C**; similarly for Theorems 9 and 10 and arbitrary colimits.

18 Let $\mathbf{C} = \mathbf{Mon}$. Prove that a monoid congruence as in the discussion preceding 3.2.4 is the same thing as a **C**-congruence.

6.3 EXAMPLES FROM AUTOMATA THEORY

The purpose of this section is to make contact with what, in recent years, has become one of the major areas of application of category theory: the mathematical study of computation and control. It is not the function of this primer to present the general categorical theory of computation and control. Rather, in the spirit of Chapter 1, we shall show in this section, and in Section 7.2, how crucial ideas of automata theory can be expressed arrow-theoretically. This should make it clear to the interested reader that the basis is laid for a general theory, just as Chapter 1 led directly to the general definitions of Chapter 2.

Imagine a machine (like one of the boxes inside a digital computer) which can be in any one of a finite number of states, receive any one of a finite number of inputs, and emit any one of a finite number of outputs. We think of it as receiving inputs, changing states, and emitting outputs once every 'cycle' of some clock which times its activity. We may represent such a system as follows:

1 **DEFINITION**: A **sequential machine** is a sextuple

$$M = (X_0, Q, \delta, q_0, Y, \beta)$$

where

X_0	is the set of **inputs**
Q	is the set of **states**
$\delta : Q \times X_0 \longrightarrow Q$	is the **dynamics** (next-state function; transition function)
$q_0 \in Q$	is the **initial state**
Y	is the set of **outputs**
$\beta : Q \longrightarrow Y$	is the **output map.**

We say M is **finite** if X_0, Q and Y are all finite sets.

Taking the 'cycle time' of M's 'clock' as the unit on our time-scale, we imagine M as representing a system which starts in state q_0 at time 0, and which is such that if it is in state $q(t) \in Q$ at time t, and then receives input $x(t)$, it will emit output

$$y(t) = \beta(q(t)) \in Y$$

and then settle into state

$$q(t+1) = \delta(q(t), x(t)) \in Q$$

by time $t+1$.

2 **EXAMPLE**: We design a sequential machine M which will add two decimal numbers in real time, so long as they are presented low-order digit first. In other words, if D is the set of decimal digits, we want to take $X_0 = D \times D$, so that at any time M will receive corresponding digits of the addends; while we take

$Y = D$, so that at any time, M will emit a digit of the sum. If we start M in state q_0, which we shall stipulate has $\beta(q_0) = 0$, we might expect the following typical behavior (noting that $372 + 41 = 413$; and that we 'feed in' numbers backwards):

Time	0	1	2	3	4	5
1st Input	2	7	3	0	0	0
2nd Input	1	4	0	0	0	0
Output	0	3	1	4	0	0
State	q_0					

Can we define Q, δ and β to make M behave this way? The reader's mental arithmetic has probably already provided the answer. "2 and 1 is 3; so put out 3, no carry"; "7 and 4 is 11, so put out 1, carry 1"; "1 and 3 and 0 is 4, so put out 4, no carry."

In other words, the state is the current output digit, together with the carry bit, which is 0 or 1. Thus we may complete the formal specification of M as follows:

$$\text{Set} \quad Q = \{0, 1\} \times D \qquad \text{with} \quad q_0 = (0, 0).$$

[Note that we may view the element (c, d) of Q as the decimal number $cd = 10c + d$ without confusion, and so we can write cd or (c, d) interchangeably.]

Then we define

$$\beta : Q \longrightarrow Y = \{0, 1\} \times D \longrightarrow D : cd \mapsto d$$

and the dynamics is given by

$$\delta : Q \times X \longrightarrow Q : \{0,1\} \times D \times D \times D \longrightarrow \{0,1\} \times D : (cd, x_1, x_2) \mapsto c'd' = c + x_1 + x_2$$

The reader should use this expression for δ to fill in the state-row of the above table, and check that β does indeed yield the stipulated outputs. The verification that M handles all input strings correctly is left as an exercise.

We shall be interested in categories for which the objects are sequential machines, and the morphisms are . . . what? To get a feel for this, suppose that, in the above example, we had not realized that the carry would always be 0 or 1, and so had formed M' with state set $Q' = D \times D$ instead of $Q = \{0, 1\} \times D$. It is clear that we can define a map h from Q to Q' by sending $(c, d) \in Q$ to $(c, d) \in Q'$. If β' and δ' are defined as are δ and β in 2, save that we always have D in place of $\{0, 1\}$, and if we take $q_0{}'$ to be 00 again, we see that the following diagrams commute:

$$\begin{array}{ccc} & \overset{q_0}{\nearrow} & Q \\ 1 & & \big\downarrow h \\ & \underset{q_0{}'}{\searrow} & Q' \end{array} \qquad \begin{array}{ccc} Q \times X_0 & \xrightarrow{\delta} & Q \\ {\scriptstyle h \times X_0}\big\downarrow & & \big\downarrow{\scriptstyle h} \\ Q' \times X_0 & \xrightarrow[\delta']{} & Q' \end{array} \qquad \begin{array}{ccc} Q & \overset{\beta}{\searrow} & \\ {\scriptstyle h}\big\downarrow & & Y \\ Q' & \underset{\beta'}{\nearrow} & \end{array}$$

where we have identified q_0 in Q with the map from a one-element set 1 to Q which has $\{q_0\}$ for its image; and where $h \times X_0$ is the map

$$h \times X_0 : Q \times X_0 \longrightarrow Q' \times X_0 : (q, x) \mapsto (h(q), x)$$

so that our three diagrams simply say

$$h(q_0) = q_0{}'$$

$$h[\delta(q, x)] = \delta'(h(q), x)) \qquad \text{for all } q \text{ in } Q \text{ and } x \text{ in } X_0$$

$$\beta(q) = \beta'(h(q)) \qquad \text{for all } q \text{ in } Q$$

which the reader can see to hold in the present example.

This has led to the introduction of two categories (among others) into the study of sequential machines, which we present in 3 and 4.

3 **DEFINITION**: Fix an input set X_0. Then the category **Dyn**(X_0) of X_0**-dynamics** has as objects

$$\text{maps } \delta : Q \times X_0 \longrightarrow Q \qquad \text{(the state-set } Q \text{ is arbitrary)}$$

which we shall usually write (Q, δ). Its morphisms $h : (Q, \delta) \longrightarrow (Q', \delta')$ are maps $h : Q \longrightarrow Q'$ which satisfy the commutative diagram

$$\begin{array}{ccc} Q \times X_0 & \xrightarrow{\delta} & Q \\ {\scriptstyle h \times X_0}\big\downarrow & & \big\downarrow{\scriptstyle h} \\ Q' \times X_0 & \xrightarrow[\delta']{} & Q' \end{array}$$

We call such an h an X_0**-dynamorphism**; and we compose dynamorphisms at the level of **Set**.

Although it is easy to check directly that **Dyn**(X_0) is a category, we shall proceed here by showing that it is a category of sets with structure:

For each set Q, the set of structures **Dyn**(X_0)(Q) is to be the set of dynamics $\delta : Q \times X \longrightarrow Q$. Then to say that

$$h : (Q, \delta) \longrightarrow (Q', \delta')$$

is admissible is just to say that h is a dynamorphism.

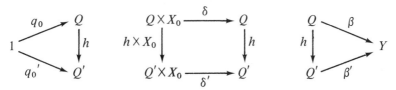

We must now check that admissible maps are composable, i.e., given dynamorphisms $h : (Q, \delta) \longrightarrow (Q', \delta')$ and $h' : (Q', \delta') \longrightarrow (Q'', \delta'')$ we must check that their composite $h' \cdot h$ is a dynamorphism. To see that it is, consider the following diagram:

$$
\begin{array}{ccccc}
& Q \times X_0 & \xrightarrow{\ \delta\ } & Q & \\
& {\scriptstyle h \times X_0}\downarrow & (B) & \downarrow{\scriptstyle h} & \\
(h' \cdot h) \times X_0 \quad (A) & Q' \times X_0 & \xrightarrow{\ \delta'\ } & Q' \quad (D) & h' \cdot h \\
& {\scriptstyle h' \times X_0}\downarrow & (C) & \downarrow{\scriptstyle h'} & \\
& Q'' \times X_0 & \xrightarrow{\ \delta''\ } & Q'' &
\end{array}
$$

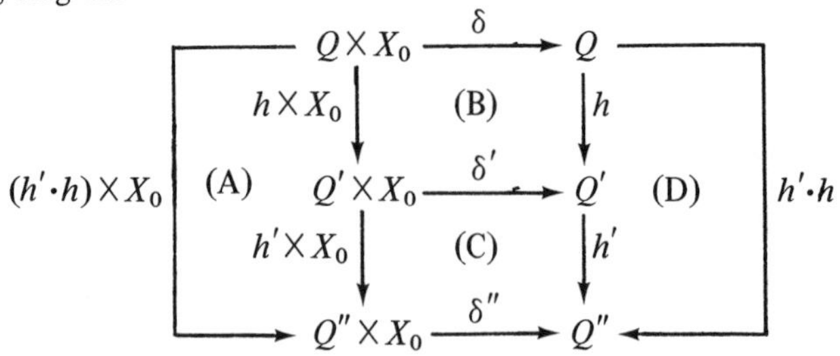

(A) commutes by the way we define $f \times X_0$ for any f:
Given any q in Q and x in X_0, we have

$$
\begin{aligned}
[(h' \cdot h) \times X_0](q, x) &= ((h' \cdot h)(q), x) \\
&= (h'(h(q)), x) \\
&= (h' \times X_0)(h(q), x) \\
&= [(h' \times X_0) \cdot (h \times X_0)](q, x).
\end{aligned}
$$

(B) and (C) commute because h and h', respectively, are dynamorphisms, while
(D) is just the definition of $h' \cdot h$. Thus we have the commutativity of the
perimeter:

$$
\begin{array}{ccc}
Q \times X_0 & \xrightarrow{\ \delta\ } & Q \\
{\scriptstyle (h' \cdot h) \times X_0}\downarrow & & \downarrow{\scriptstyle h' \cdot h} \\
Q'' \times X_0 & \xrightarrow[\ \delta''\ ]{} & Q''
\end{array}
$$

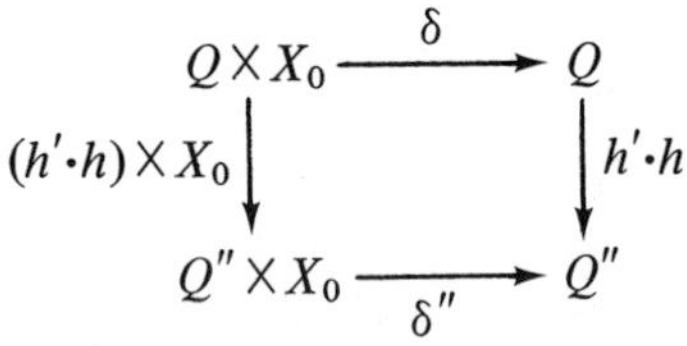

so that $h' \cdot h$ is indeed admissible, $h' \cdot h : (Q, \delta) \rightarrow (Q'', \delta'')$.

Finally we must check, given a dynamics (Q, δ) and a bijection $h : Q \rightarrow Q'$
that there exists a unique dynamics δ' such that $h : (Q, \delta) \rightarrow (Q', \delta')$ and
$h^{-1} : (Q', \delta') \rightarrow (Q, \delta)$ are admissible. But the admissibility diagrams

$$
\begin{array}{ccc}
Q \times X_0 & \xrightarrow{\ \delta\ } & Q \\
{\scriptstyle h \times X_0}\downarrow & & \downarrow{\scriptstyle h} \\
Q' \times X_0 & \dashrightarrow{\ \delta'\ } & Q' \\
{\scriptstyle h^{-1} \times X_0}\downarrow & & \downarrow{\scriptstyle h^{-1}} \\
Q \times X_0 & \xrightarrow[\ \delta\ ]{} & Q'
\end{array}
$$

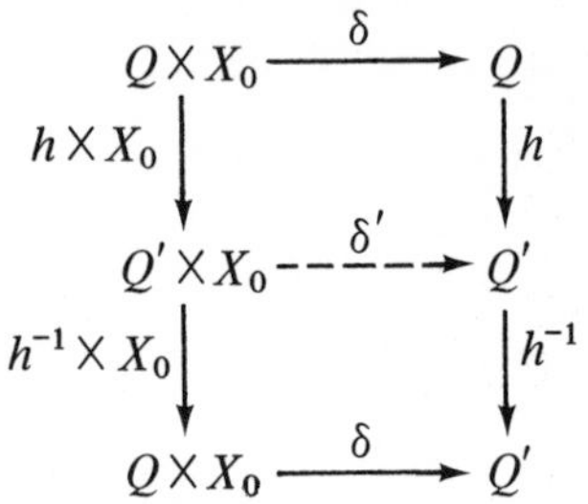

(in which the bottom square is essentially the same as the top one!) are uniquely
satisfied by setting $\delta'(q', x) = h\delta(h^{-1}(q'), x)$ for each q' in Q' and x in X_0.

Thus $\mathbf{Dyn}(X_0)$ is indeed a category of sets with structure and so, *a fortiori*, a category.

We now build on 3 to define a category of machines:

4 **DEFINITION**: Fix an input set X_0 and an output set Y. The category **Mach**(X_0, Y) of (X_0, Y)-machines has as objects

sequential machines $M = (X_0, Q, \delta, q_0, Y, \beta)$ (the state-set Q is arbitrary).

A morphism $h : M \to M'$ is a dynamorphism $h : (Q, \delta) \to (Q', \delta')$ which satisfies the additional requirement that

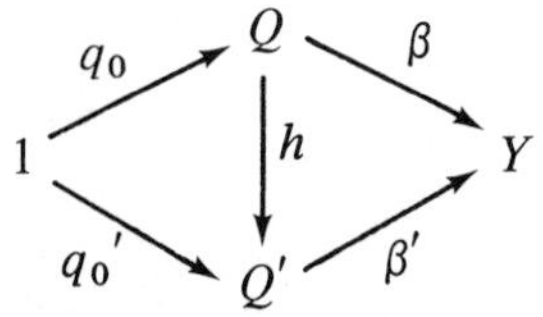

commutes. We call such an h a **simulation**, and say that M simulates M'. Composition of simulations is at the level of **Set**.

We leave to the reader the straightforward verification that the composition of simulations is again a simulation, etc. [exercise 2].

Two questions of importance to automata theorists (and control theorists) are the following:

(1) Given a state q of a sequential machine (or control system) M does there exist an input sequence (control sequence) which will drive M from its initial state q_0 to the desired state q?

(2) Given a machine (control system) M which may be in either state q_1 or q_2, does there exist an input sequence, M's response to which will enable us to tell the two states apart?

These are the problems of **reachability** and **observability**, and it is to their formal study that we now turn:

We know that a single input x in X_0 will send M from state q to state $\delta(q, x)$, from which a second input x' will send M to state $\delta(\delta(q, x), x')$, from which a third input x'' will send M to state $\delta(\delta(\delta(q, x), x'), x'')$, and so on. In this way, we may specify the state-behavior of M in response to any sequence w of inputs, i.e. any element of the free monoid X_0^* on X_0 generators:

5 **DEFINITION**: Given an X_0-dynamics (Q, δ), we define its **run map** to be the unique map

$$\delta^* : Q \times X_0^* \to Q$$

defined inductively by

Basis Step: $\delta^*(q, \Lambda) = q$ for all q in Q

Induction Step: $\delta^*(q, wx) = \delta(\delta^*(q, w), x)$ for all q in Q, w in X_0^*, x in X.

Note that, looking at an input x in X_0 as a one-element string (x) in X_0^*, we do indeed have

$$\delta^*(q, (x)) = \delta(\delta^*(q, \Lambda), x) = \delta(q, x)$$

as desired.

Now the basis step may be rewritten as the commutative diagram

$$\begin{array}{ccc} Q & \xrightarrow{\eta Q} & Q \times X_0{}^* \\ & \searrow^{\mathrm{id}_Q} & \downarrow{\scriptstyle \delta^*} \\ & & Q \end{array} \tag{1}$$

where $\eta Q : Q \longrightarrow Q \times X_0{}^*$ is simply the map $q \mapsto (q, \Lambda)$. Again, the induction step may be rewritten as

$$\begin{array}{ccc} (Q \times X_0{}^*) \times X_0 & \xrightarrow{\mu_0 Q} & Q \times X_0{}^* \\ {\scriptstyle \delta^* \times X_0}\downarrow & & \downarrow{\scriptstyle \delta^*} \\ Q \times X_0 & \xrightarrow[\delta]{} & Q \end{array} \tag{2}$$

where $\mu_0 Q : (Q \times X_0{}^*) \times X_0 \longrightarrow Q \times X_0{}^* \times X_0$ is simply the map $((q, w), x) \mapsto (q, wx)$. In other words, $(Q \times X_0{}^*, \mu_0 Q)$ is a dynamics – we call it the **free X_0-dynamics on Q generators** – and we have:

6 **FACT**: The run map of a dynamics (Q, δ) is the unique dynamorphism $\delta^* : (Q \times X_0{}^*, \mu_0 Q) \longrightarrow (Q, \delta)$ which satisfies the condition

$$\delta^* \cdot \eta Q = \mathrm{id}_Q \, . \qquad \square$$

If we now make the definition:

7 **DEFINITION**: The **reachability** map of a dynamics (Q, δ) with initial state q_0 is the map

$$r : X_0{}^* \longrightarrow Q : w \mapsto \delta^*(q_0, x).$$

we may equally well view $X_0{}^*$ as its product $1 \times X_0{}^*$ with a one-element set, and define r by the diagram

$$r : 1 \times X_0{}^* \xrightarrow{q_0 \times X_0{}^*} Q \times X_0{}^* \xrightarrow{\delta^*} X_0{}^*$$

[where $(q_0 \times X_0{}^*)(1, w) = (q_0, w)$] and we may use (1) and (2) to obtain the commutative diagrams

$$\begin{array}{ccc} 1 & \xrightarrow{\eta 1} & 1 \times X_0{}^* \\ & \searrow^{q_0} & \downarrow{\scriptstyle r} \\ & & Q \end{array} \qquad \begin{array}{ccc} (1 \times X_0{}^*) \times X_0 & \xrightarrow{\mu_0 1} & 1 \times X_0{}^* \\ {\scriptstyle r \times X_0}\downarrow & & \downarrow{\scriptstyle r} \\ Q \times X_0 & \xrightarrow[\delta]{} & Q \end{array} \tag{3}$$

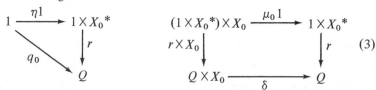

so that (the reader should check the details)

8 FACT: The reachability map of a dynamics (Q, δ) with initial state q_0 is the unique dynamorphism $r : (1 \times X_0{}^*, \mu_0 1) \longrightarrow (Q, \delta)$ which satisfies the condition

$$r \cdot \eta 1 = q_0 . \qquad \square$$

Note that $r \cdot \eta 1 = q_0$ just says $r(\Lambda) = q_0$.

Thus we have shown that two key concepts of automata theory may be defined purely in arrow terms, as in the characterizations of δ^* and r in 7 and 8. Thus, they 'live' in **Dyn**(X_0).

We close this initial discussion of reachability by saying that M is *reachable* just in case every q in Q is an $r(w) = \delta^*(q_0, w)$ for some input sequence in $X_0{}^*$. In categorical terms, this becomes:

9 DEFINITION: A dynamics (Q, δ) with initial state q_0 is said to be **reachable** just in case its reachability map is an epimorphism.

We now formalize observability: We want to see how M responds when started in some state q. Now, when M is in any state q', it emits output $\beta(q')$. Thus if we start M in state q and apply input sequence w, then M will go to state $\delta^*(q, w)$ and hence emit output $\beta[\delta^*(q, w)]$. Formally:

10 DEFINITION: Given an X_0-dynamics (Q, δ) with output map $\beta : Q \to Y$, we define the **response** (or **behavior**) **of** (Q, δ, β) **started in state** q to be

$$M_q : X_0{}^* \longrightarrow Y : w \mapsto \beta(\delta^*(q, w)).$$

Thus M_q is an element of the set $Y^{X_0{}^*}$ of all functions from $X_0{}^*$ to Y. We now see how the behavior changes as we go from q to $\delta(q, x)$:

$$\begin{aligned} M_{\delta(q, x)}(w) &= \beta(\delta^*(\delta(q, x), w)) && \text{by definition of } M_{\delta(q, x)} \\ &= \beta(\delta^*(q, xw)) && \text{by the inductive definition of } \delta^* \\ &= M_q(xw) . \end{aligned}$$

Thus if we define the map $L_x : X_0{}^* \longrightarrow X_0{}^*$ to be the **left translation** $w \mapsto xw$, we have that

$$M_{\delta(q, x)} = M_q L_x .$$

In other words, if we let $\sigma : Q \longrightarrow Y^{X_0{}^*}$ be the **observability map** which assigns M_q to each q in Q, we have the commutativity of

$$\begin{array}{ccc} Y^{X_0{}^*} \times X_0 & \xrightarrow{LY} & Y^{X_0{}^*} \\ {\scriptstyle \sigma \times X_0} \uparrow & & \uparrow {\scriptstyle \sigma} \\ Q \times X_0 & \xrightarrow[\delta]{} & Q \end{array} \qquad (4)$$

where $LY : Y^{X_0^*} \times X_0 \longrightarrow Y^{X_0^*}$ is simply the map $(f, x) \mapsto f \cdot L_x$. In other words, $(Y^{X_0^*}, LY)$ is a dynamics – we call it the **cofree dynamics on** Y **generators** – and σ is a dynamorphism $(Q, \delta) \longrightarrow (Y^{X_0^*}, LY)$.

Of course, besides satisfying this diagram, σ must satisfy the equality

$$\sigma(q)(\Lambda) = M_q(\Lambda) = \beta(\delta^*(q, \Lambda)) = \beta(q)$$

so that we have the commutativity of

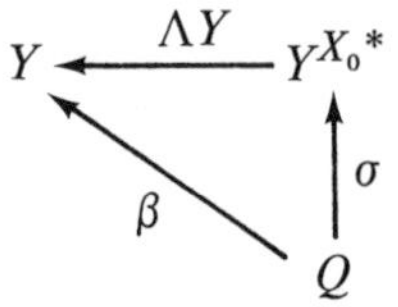

(5)

where $\Lambda Y : Y^{X_0^*} \longrightarrow Y : f \mapsto f(\Lambda)$. (The two-fold use of Λ for the empty string, and for **evaluation** at the empty string should cause no confusion.) It is straightforward (exercise 4) to check that $q \mapsto M_q$ is the only σ which satisfies (4) and (5), and we thus have (analogously to 6):

11 FACT: The observability map of a dynamics (Q, δ) with output map $\beta : Q \longrightarrow Y$ is the unique dynamorphism $\sigma : (Q, \delta) \longrightarrow (Y^{X_0^*}, LY)$ which satisfies the condition

$$\Lambda Y \cdot \sigma = \beta. \qquad \square$$

Now we can distinguish state q_1 from state q_2 just in case there exists an input sequence w such that $M_{q_1}(w) \neq M_{q_2}(w)$, i.e. just in case M_{q_1} and M_{q_2} are distinct elements of $Y^{X_0^*}$. Thus all pairs of states are distinguishable just in case $\sigma : q \mapsto M_q$ is one-to-one. In categorical terms, this becomes:

12 DEFINITION: A dynamics (Q, δ) with output map $\beta : Q \longrightarrow Y$ is said to be **observable** just in case its observability map is a monomorphism.

Inspection of 9 and 12 may suggest to the astute reader that reachability and observability are *dual* concepts, in the categorical sense of the term. This is indeed so – and for a class of machines broader than the sequential machines we study here – but the details would take us beyond the level of this introductory treatment. In such a general treatment, we replace epimorphisms and monomorphisms by the more general **E** and **M** of an image factorization system (2.3.13). Instead, we close this section by showing how to apply the image factorization theory of Section 2.3 to the automata theory problem of minimal realization:

13 EXAMPLE: Consider the finite sequential machine M represented by the following *state graph*, with states q_0, q_1, q_2, where q/y means $y = \beta(q)$ and where $\delta(q, x)$ is the unique terminus of the arrow labelled x emanating from q:

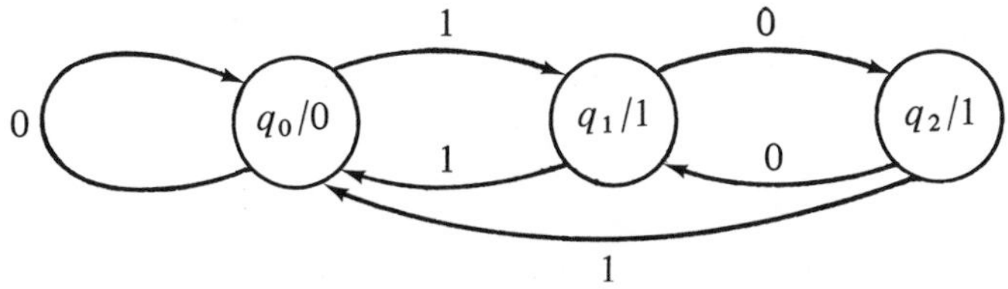

We may partition its state set into two pieces, $Q_0 = \{q_0\}$ and $Q_1 = \{q_1, q_2\}$, states of Q_0 having output 0, and states of Q_1 having output 1. We see that M stays in either Q_j so long as the input is 0, but moves from Q_j to Q_{1-j} each time the input is 1. It is thus clear that M_{q_0} is just the function

$$p(w) = \begin{cases} 0 & \text{if } w \text{ contains an even number of 1's} \\ 1 & \text{if } w \text{ contains an odd number of 1's} \end{cases}$$

while for $q = q_1$ or q_2, we have M_q to be

$$\bar{p}(w) = \begin{cases} 1 & \text{if } w \text{ contains an even number of 1's} \\ 0 & \text{if } w \text{ contains an odd number of 1's.} \end{cases}$$

Thus we may think of M, started in state q_0, to be a **parity checker.** It is also clear that, to build a parity checker – i.e. a sequential machine M' whose initial-state response is p (we say M' is a **realization** of p) – we do not need all three states of M. In fact, we may 'merge' q_1 and q_2 to obtain the two-state realization of p:

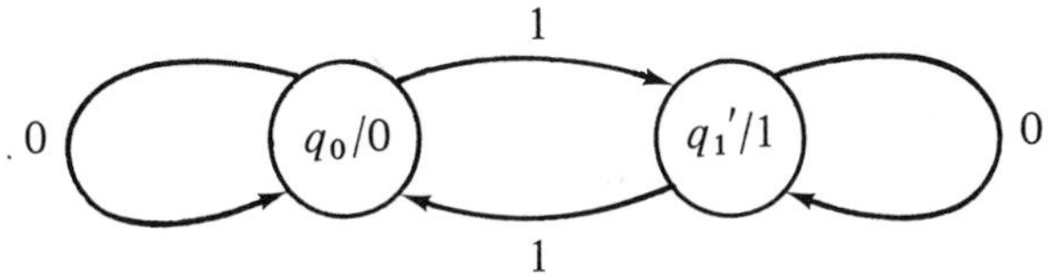

(6)

and it is clear that no one-state machine could do the job since it could only ever emit one value of the output. Thus we may say that the M' of (6) is the **minimal realization** of p in that it has the smallest number of states of any realization of p. Our task now is to show that we can characterize M' in categorical terms:

Let, then, $M = (\{0, 1\}, Q, \delta, q_0, \{0, 1\}, \beta)$ be any *reachable* realization of p. That is, we have

(i) $p = M_{q_0}$ so that $\beta(\delta^*(q_0, w)) = p(w)$ for all w in X_0^*; and

(ii) $r : X_0^* \longrightarrow Q$ is an epimorphism, so that every q in Q equals $\delta^*(q_0, w)$ for some w in X_0^*.

Now recall from our discussion following 10 that

$$M_{\delta(q,x)} = M_q L_x \qquad \text{for each } q \text{ in } Q \text{ and } x \text{ in } X_0.$$

If we now define $L_{w'} : X_0{}^* \longrightarrow X_0{}^* : w \mapsto w'w$ for any w' in $X_0{}^*$ (so that L_Λ is just $\mathrm{id}_{X_0{}^*}$), it is straightforward induction to verify that

$$M_{\delta^*(q,\, w')} = M_q L_{w'} \qquad \text{for each } q \text{ in } Q \text{ and } w' \text{ in } X_0{}^*.$$

But from this we immediately deduce that

$$M_{\delta^*(q_0,\, w')} = \begin{cases} p & \text{if } w' \text{ contains an even number of 1's} \\ \bar{p} & \text{if } w' \text{ contains an odd number of 1's} \end{cases}$$

since

$$M_{\delta^*(q_0,\, w')}(w) = M_{q_0}(w'w) = p(w'w).$$

Thus the responses of the states of *any* reachable realization M of p correspond to the states of the minimal realization M' of (6), and we have:

14 OBSERVATION: Let M be a reachable realization of p, and let M' be the minimal realization of (6). Then the map

$$h : Q \longrightarrow \{q_0{}', q_1{}'\} : r(w) \mapsto \text{the } q' \text{ such that } M_{r(w)} = M'_{q'}$$

(note the essential use made in defining h of the fact that r is onto) is a *simulation*; and is in fact the unique simulation of M' by M.

Proof: Clearly $h(q_0) = q_0{}'$. To see that h is a dynamorphism, we have

$$\begin{aligned} h(\delta(r(w), x)) &= h(r(wx)) \\ &= \text{the } q'' \text{ such that } M_{r(wx)} = M'_{q''} \\ &= \delta(h(r(w)), x) \end{aligned}$$

since $M'_{\delta(q',\, x)} = M'_{q''}$ iff $\delta'(q', x) = q''$.

The final check that h is a simulation is to note that for any q in Q,

$$\beta(q) = M_q(\Lambda) = \beta'(h(q)).$$

But this last equation also makes clear that the choice of h is unique, for it reduces to $h(q) = M'_{\beta(q)}$. □

Now whereas the construction of (6), and the criterion of "minimal number of states" was all very *ad hoc* and uncategorical, the criterion given in 14 is indeed categorical and we are now ready to give the theory of minimal realization in algebraic form:

15 DEFINITION: Let $M = (X_0, Q, \delta, q_0, Y, \beta)$ be an (X_0, Y)-machine, and let its reachability map be $r : X_0{}^* \longrightarrow Q$, and its observability map be $\sigma : Q \longrightarrow Y^{X_0{}^*}$. Then its **total response** map is the composition

$$f^\blacktriangle = \sigma \cdot r : X_0{}^* \longrightarrow Y^{X_0{}^*}.$$

We say M is a realization of some $g : X_0{}^* \longrightarrow Y^{X_0{}^*}$ just in case g is M's total response map.

Note that since $r : (X_0^*, 1\mu_0) \longrightarrow (Q, \delta)$ and $\sigma : (Q, \delta) \longrightarrow (Y^{X_0^*}, LY)$ are both dynamorphisms, it follows that $f^\blacktriangle : (X_0^*, 1\mu_0) \longrightarrow (Y^{X_0^*}, LY)$ is also a dynamorphism. Note that $f^\blacktriangle(w')$ is in $Y^{X_0^*}$ and that

$$[f^\blacktriangle(w')](w) = M_{q_0}(w'w).$$

Thus $M_{q_0} = f^\blacktriangle(\Lambda)$, so that each of $f^\blacktriangle$ and M_{q_0} determines the other. In the general theory below, it will be more convenient to consider realizations of $f^\blacktriangle$, rather than to work with M_{q_0} as in 13.

16 **OBSERVATION**: Let $h : M \longrightarrow M'$ be a simulation in **Mach**(X_0, Y), i.e. $h : (Q, \delta) \longrightarrow (Q', \delta')$ is a dynamorphism satisfying

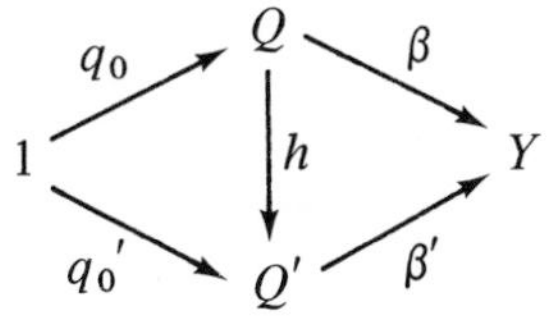

Then M and M' have the same total response map.

Proof: Exercise 5. □

Then Observation 14 immediately suggests the general definition:

17 **DEFINITION**: Given a dynamorphism $f^\blacktriangle : (X_0^*, \mu_0 1) \longrightarrow (Y^{X_0^*}, LY)$, we say that a realization M' is **minimal** if it is reachable, and if for *any* reachable realization M of $f^\blacktriangle$ there exists a *unique* simulation $M \longrightarrow M'$; i.e. just in case M' is *terminal* in the subcategory **RR**$(f^\blacktriangle)$ of **Mach**(X_0, Y) whose objects are the reachable realizations of $f^\blacktriangle$, and whose morphisms are the simulations.

Our new definition thus yields without any effort the result:

18 **OBSERVATION**: The minimal realization of $f^\blacktriangle$ is unique up to isomorphism, i.e. if M and M' are both minimal realizations of $f^\blacktriangle$, then there exist unique simulations $\psi : M \longrightarrow M'$ and $\psi' : M' \longrightarrow M$ such that $\psi' \cdot \psi = \mathrm{id}_Q$ and $\psi \cdot \psi' = \mathrm{id}_{Q'}$, so that M and M' differ only in the 'labelling' of their states. □

Now let's reexamine the construction of the minimal realization M' of p in 13. We proved that

$$M_{\delta^*(q, w')} = \begin{cases} p & \text{if } w' \text{ contains an even number of 1's} \\ \bar{p} & \text{if not.} \end{cases}$$

In other words, the *image* of $p^\blacktriangle : \{0, 1\}^* \longrightarrow \{0, 1\}^{\{0, 1\}^*}$ is just $\{p, \bar{p}\}$. The way we constructed $h : Q \longrightarrow \{q_0', q_1'\}$ may be viewed as a map

$$X_0^* \longrightarrow \mathrm{Im}(p^\blacktriangle) : w \mapsto p^\blacktriangle(w)$$

where we work directly with w rather than the state $r(w)$ of the particular realization M, and where we identify each state of M' with its response. In the remainder of this section, we shall show that this situation holds in general: given any dynamorphism $f^\blacktriangle$, its image factorization yields the state-space $\mathrm{Im}(f^\blacktriangle)$ of a minimal realization, and the minimal realization is not only reachable but also observable. We start by showing how to put a dynamic structure on the image of *any* dynamorphism:

19 **DYNAMORPHIC IMAGE LEMMA:** Let $h : (Q, \delta) \longrightarrow (Q', \delta')$ be an X_0-dynamorphism, and let (e, m) be an image factorization of h. Then there exists a unique dynamical structure δ'' on $h(Q)$ such that $e : (Q, \delta) \longrightarrow (h(Q), \delta'')$ and $m : (h(Q), \delta'') \longrightarrow (Q', \delta')$ are dynamorphisms.

Proof: It is clear that if $e : Q \longrightarrow h(Q)$ is an epimorphism, then so too is $e \times X_0 : Q \times X_0 \longrightarrow h(Q) \times X_0 : (q, x) \mapsto (h(q), x)$. Thus we may use diagonal fill-in (2.3.14) to find the unique δ'' such that

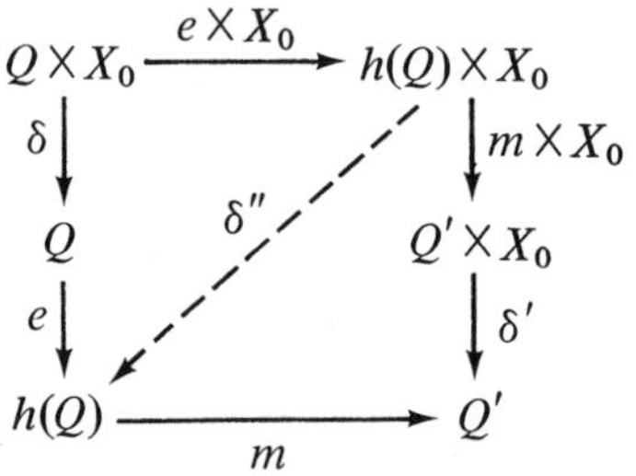

commutes. (Note that the outer square simply says $m \cdot e = h$ is a dynamorphism, and so commutes.) But the upper triangle says $e : (Q, \delta) \longrightarrow (h(Q), \delta'')$ is a dynamorphism, while the lower triangle says that $m : (h(Q), \delta'') \longrightarrow (Q', \delta')$ is a dynamorphism. □

Incidentally, the above proof uses only the properties of image factorization systems, and so conveys some flavor of how easily the current approach to automata theory generalizes. We now introduce canonical realizations and note a useful property, 21, after which we can prove the posited properties of minimal realizations:

20 **DEFINITION:** A realization of a total response map is **canonical** if it is both reachable and observable.

21 **SIMULATION LEMMA:** Let M be a reachable realization of a total response map $f^\blacktriangle$, and let M' be an observable realization of $f^\blacktriangle$. Then there exists a unique simulation $h : M \longrightarrow M'$.

Proof: The reachability map r of M is an epimorphism, while the observability map σ' of M' is a monomorphism. Thus by diagonal fill-in, there exists a unique

h such that

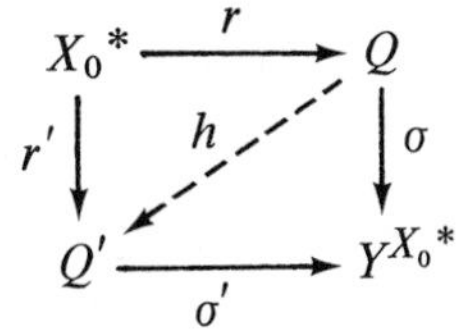

commutes. Since $h \cdot r = r'$ with r and r' dynamorphisms, and since r is an epimorphism, it follows (by exercise 6) that h is the required dynamorphism by which M simulates M'. □

In particular, this says that any canonical realization of $f^\blacktriangle$ (if it has any) is minimal. In fact, we can prove much more:

22 **MINIMAL REALIZATION THEOREM**: Every dynamorphism $f^\blacktriangle : (X_0^*, \mu_0 1) \longrightarrow (Y^{X_0^*}, LY)$ has a minimal realization M_f. A system M is a minimal realization of $f^\blacktriangle$ if and only if M is a canonical realization of $f^\blacktriangle$.

Proof: Let (r_f, σ_f) be an image factorization

$$X_0^* \xrightarrow{r_f} Q_f \xrightarrow{\sigma_f} Y^{X_0^*}$$

of $f^\blacktriangle$. By the dynamorphic image lemma, there exists a unique dynamics $\delta_f : Q_f \times X_0 \longrightarrow Q_f$ such that $r_f : (X_0^*, \mu_0 1) \longrightarrow (Q, \delta)$ and $\sigma_f : (Q_f, \delta_f) \longrightarrow (Y^{X_0^*}, LY)$ are both dynamorphisms. But if we define $q_f \in Q_f$ and $\beta_f : Q_f \longrightarrow Y$ by the diagrams

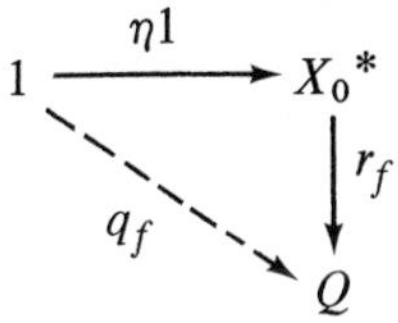

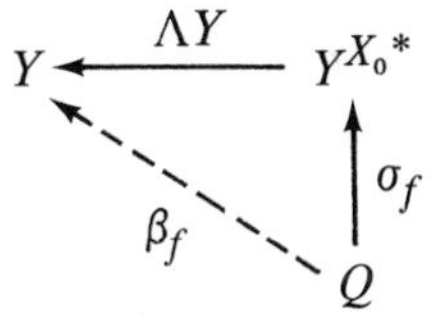

it is then immediate from 8 and 11 that the sequential machine

$$M_f = (X_0, Q_f, \delta_f, q_f, Y, \beta_f)$$

has reachability map r_f and observability map σ_f, and hence has total response map $f^\blacktriangle = \sigma_f \cdot r_f$. By the nature of image factorizations, r_f is epi and σ_f is mono, so that M_f is a canonical – and hence minimal – realization of $f^\blacktriangle$.

Since all minimal realizations of $f^\blacktriangle$ are isomorphic to M_f, they are certainly canonical, and we have already see that canonical realizations are minimal. □

With this we conclude our rendering of sequential machine theory into arrow form – save that in Section 7.2 we shall see that our definitions of free and cofree dynamics above are not *ad hoc* but in fact follow from the general notion of adjoints of functors.

Before closing this section, however, a comment may be necessary for the reader, already expert in automata theory, who finds our arrow treatment of observability, reachability and minimal realization longer than the usual element-by-element treatment with which he is familiar. There are two immediate rejoinders: Firstly, since we have presented most of the element-by-element theory as motivation for the arrow treatment, much of the present bulk is 'padding' with respect to the categorical treatment. Secondly, since diagram-chasing is new to most readers, the arrow treatment is heavier going for the novice than it will be when you have become an adept "commuter". However, the real answer – which should by now be fairly apparent to the reader who has seen the drastically different form taken by coproducts, say, in the categories **Set**, **Vect** and **Grp** – is that our arrow definitions and diagram-chasing proofs generalize *without significant further effort* to many classes of machines – such as the linear discrete-time systems of control theory – other than sequential machines. The element-by-element definitions and proofs do not.

Exercises

1 Take the machine M defined in 2. Show that if it is started in state $q_0 = 00$ at time 0, and receives the input string

$$(x_{01}, x_{02}), (x_{11}, x_{12}), ..., (x_{n1}, x_{n2}), (0, 0)(0, 0)$$

where input $(x_{t1}, x_{t2}) \in X_0$ is applied at time t, then the resultant output string will be

$$0y_1y_2 \cdots y_{n+1}y_{n+2}$$

(with $y_t \in Y$ emitted at time t) where the y's are defined by the equation

$$y_{n+2} \cdots y_2 y_1 = x_{n1} \cdots x_{11} x_{01} + x_{n2} \cdots x_{12} x_{02}$$

where the strings of digits are now interpreted as decimal numbers, and + denotes ordinary numerical addition.

2 Verify that **Mach**(X_0, Y) is a category of sets with structure.

3 Check the commutativity of (3) and prove Fact 8.

4 Verify that (4) and (5) define σ uniquely in terms of δ and β.

5 Verify Observation 16. [Hint: Prove that $h \cdot r = r'$ and $\sigma = \sigma' \cdot h$.]

6 Let $f: Q_1 \longrightarrow Q_2$ and $g: Q_2 \longrightarrow Q_3$ be such that $f: (Q_1, \delta_1) \longrightarrow (Q_2, \delta_2)$ and $g \cdot f: (Q_1, \delta_1) \longrightarrow (Q_3, \delta_3)$ are both dynamorphisms. Show that if f is an epimorphism then g is a dynamorphism $(Q_2, \delta_2) \longrightarrow (Q_3, \delta_3)$. That is, show that surjective dynamorphisms are co-optimal.

Chapter 7

FUNCTORS AND ADJOINTS

With this chapter, we begin the second part of our book, in which we build upon our study of arrows and structure to explore the third facet of the categorical imperative: *functors.*

7.1 FUNCTORS

We have seen that a one-object category is a monoid (3.1.3); let us now see a sense in which any category is a *generalized* monoid. Given **K**, let

$$\mathbf{M_K} = \coprod_{A,B \in \mathrm{Obj}(\mathbf{K})} \mathbf{K}(A, B)$$

be the collection of all **K**-morphisms. Composition defines a *partial* function

$$\mathbf{M_K} \times \mathbf{M_K} \longrightarrow \mathbf{M_K} : (f, g) \mapsto g \cdot f.$$

Let us see how we may recapture the identities of **K** from this function: We say that u is an identity if $g \cdot u = g$ whenever $g \cdot u$ is defined, and if $u \cdot f = f$ whenever $u \cdot f$ is defined. As before, each f has exactly one identity u such that $u \cdot f$ is defined, call it $C(f)$ for the identity of the *codomain* of f; and exactly one v such that $f \cdot v$ is defined, call it $D(f)$ for the *domain* of f. Thus our partial function $\mathbf{M_K} \times \mathbf{M_K} \longrightarrow \mathbf{M_K}$ satisfies the conditions:

1 There are total functions

$$C, D : \mathbf{M_K} \longrightarrow \text{identities of } \mathbf{M_K}$$

for which

$$D(D(f)) = D(f) = C(D(f));$$
$$D(C(f)) = C(f) = C(C(f)); \quad \text{and}$$
$$f \cdot D(f) = f = C(f) \cdot f.$$

2 $g \cdot f$ is defined iff $C(f) = D(g)$. If $g \cdot f$ is defined then $D(g \cdot f) = D(f)$ and $C(g \cdot f) = C(g)$. Moreover, if either $(h \cdot g) \cdot f$ or $h \cdot (g \cdot f)$ is defined, then both are defined and $(h \cdot g) \cdot f = h \cdot (g \cdot f)$.

[In fact, given a partial function satisfying 1 and 2 we can reconstitute a category, with objects being the identities (exercise 1).]

Generalized monoids cry out for generalized homomorphisms $H : \mathbf{M_K} \longrightarrow \mathbf{M_L}$ and the obvious demand is that

$$f \text{ an identity} \Rightarrow Hf \text{ an identity} \tag{1}$$

$$g{\cdot}f \text{ defined} \Rightarrow H(g{\cdot}f) = Hg{\cdot}Hf. \tag{2}$$

For each object A of $\mathbf{K}$ denote by HA the object on which $H(\mathrm{id}_A)$ is the identity, i.e. such that $H(\mathrm{id}_A) = \mathrm{id}_{HA}$. Since $f{\cdot}\mathrm{id}_A$ is defined for $f : A \longrightarrow B$, we have from (2) that

$$Hf = H(f{\cdot}\mathrm{id}_A) = Hf{\cdot}H(\mathrm{id}_A) = Hf{\cdot}\mathrm{id}_{AH}.$$

Hence Hf, being composable with id_{AH}, must have domain AH. Similarly, it must have codomain BH since $\mathrm{id}_B{\cdot}f$ is defined. Thus (2) says, in short, $H(A \xrightarrow{f} B) = HA \xrightarrow{Hf} HB$.

This, then, motivates the following:

3 **DEFINITION**: A **functor** H from a category $\mathbf{K}$ to a category $\mathbf{L}$ is a function which maps $\mathrm{Obj}(\mathbf{K}) \longrightarrow \mathrm{Obj}(\mathbf{L}) : A \mapsto HA$, and which for each pair A, B of objects of $\mathbf{K}$ maps $\mathbf{K}(A, B) \longrightarrow \mathbf{L}(HA, HB) : f \mapsto Hf$, while satisfying the two conditions:

$$H(\mathrm{id}_A) = \mathrm{id}_{HA} \qquad \text{for every } A \in \mathrm{Obj}(\mathbf{K})$$

$$H(g{\cdot}f) = Hg{\cdot}Hf \quad \text{whenever } g{\cdot}f \text{ is defined in } \mathbf{K}.$$

We say H is an **isomorphism** if $A \mapsto HA$ and each $\mathbf{K}(A, B) \longrightarrow \mathbf{L}(HA, HB)$ are bijections.

The most trivial example of a functor (or of an isomorphism), is, for each category $\mathbf{K}$, the **identity functor** $\mathrm{id}_{\mathbf{K}} : \mathbf{K} \longrightarrow \mathbf{K}$ which sends A to A and f to f.

A more interesting functor is the functor $-\times X_0 : \mathbf{Set} \longrightarrow \mathbf{Set}$, where X_0 is a fixed set, which sends each set Q to the set $Q \times X_0$, and sends each map $f : Q \longrightarrow Q'$ to the map $f \times X_0 : Q \times X_0 \longrightarrow Q' \times X_0 : (q, x) \mapsto (f(q), x)$. [Thus $f \times X_0$ could be written as $f \times \mathrm{id}_{X_0}$. However, as exercise 1 makes clear, we may regard an object and its identity as being formally equivalent.] To see that $-\times X_0$ is a functor note that:

$$\mathrm{id}_Q \times X_0 = \mathrm{id}_{Q \times X_0} : Q \times X_0 \longrightarrow Q \times X_0 : (q, x) \mapsto (q, x)$$

while for $f : Q \longrightarrow Q'$ and $g : Q' \longrightarrow Q''$ we have

$$(g{\cdot}f) \times X_0 = (g \times X_0){\cdot}(f \times X_0) : Q \times X_0 \longrightarrow Q'' \times X_0 : (q, x) \mapsto (g(f(q)), x).$$

Next, we give an interesting example of an isomorphism of categories.

Pfn is the category of **sets and partial functions**: given sets A and B, a morphism $f : A \longrightarrow B$ in **Pfn** is a *partial* function from A to B, i.e. a map $d(f) \longrightarrow B$

for some (possibly empty) subset $d(f)$ of A called the **domain of definition** of f. [If $d(f) = A$, we say f is **total** – i.e. just in case it is an $f : A \longrightarrow B$ in the sense of **Set**.]

Fix a one-element set $\{*\}$. We say a set is **pointed** if it is of the form $A + \{*\}$ (disjoint union) for some set A. We say a map $f : A + \{*\} \longrightarrow B + \{*\}$ is **pointed** if $f(*) = *$. Then $\mathbf{Set}_*$ is the category of pointed sets and pointed maps.

It is then clear that the assignment $H : \mathbf{Set}_* \longrightarrow \mathbf{Pfn}$ defined by

$$H(A + \{*\}) = A$$

$$Hf : A \longrightarrow B : a \mapsto \begin{cases} f(a) & \text{if } f(a) \neq * \\ \text{undefined} & \text{if not} \end{cases}$$

(so that $d(Hf) = A \backslash f^{-1}(*)$) is not only a functor but also an isomorphism of categories.

A general and important class of functors are the **forgetful functors**. For example monoids are sets with extra structure and homomorphisms are maps which satisfy extra conditions. Thus we have a well-defined mapping

$$U : \mathbf{Mon} \longrightarrow \mathbf{Set}$$

which sends a monoid to its underlying set, and forgets that a homomorphism has anything special about it. Thus $U(\mathrm{id}_M) = \mathrm{id}_M$ and $U(g \cdot f) = g \cdot f$ and so U is certainly a functor. Similarly, we can define forgetful functors $\mathbf{Grp} \longrightarrow \mathbf{Mon}$, $\mathbf{Grp} \longrightarrow \mathbf{Set}$, $\mathbf{Vect} \longrightarrow \mathbf{Grp}$, $\mathbf{Met} \longrightarrow \mathbf{Top}$, etc., etc., etc.

At first sight, the concept of a forgetful functor seems so trivial as to *appear* completely useless. Surprisingly, we shall see that this impression is completely false. For example, in the next section, we shall explore the possibility of associating to each functor G another functor F, called the left adjoint of G. We shall see that the left adjoint of the forgetful functor $U : \mathbf{Mon} \longrightarrow \mathbf{Set}$ is the functor $F : \mathbf{Set} \longrightarrow \mathbf{Mon}$ which sends each set B to the free monoid $FB = (B^*, \mathrm{conc}, \Lambda)$ on B generators. In fact, many important constructions, such as those in our discussion of sequential machines in Section 6.3, will be revealed as *adjoints of forgetful functors.*

Returning now to Definition 3, we should note that many authors call such a functor **covariant**, contrasting it with a map $H : \mathbf{K} \longrightarrow \mathbf{L}$ for which we still have $H(\mathrm{id}_A) = \mathrm{id}_{AH}$, but which reverses the order of composition: $H(f \cdot g) = Hg \cdot Hf$ whenever $f \cdot g$ is defined in $\mathbf{K}$. Such an order-reversing map is called a **contravariant** functor. However, we do not need this extra concept because it is clear that:

4 A contravariant functor $\mathbf{K} \longrightarrow \mathbf{L}$ is simply a (covariant) functor $\mathbf{K}^{\mathrm{op}} \longrightarrow \mathbf{L}$. □

We can form functors of several variables by defining product categories:

5 **DEFINITION**: Given two categories **K** and **L**, we define their **product** $\mathbf{K} \times \mathbf{L}$ to be the category whose objects are ordered pairs (K, L) of objects K from **K** and L from **L**, and for which morphisms

$$(K, L) \longrightarrow (K', L')$$

are just pairs (f, g) with $f \in \mathbf{K}(K, K')$ and $g \in \mathbf{L}(L, L')$, while

$$\mathrm{id}_{(K,L)} = (\mathrm{id}_K, \mathrm{id}_L) \quad \text{and} \quad (f', g')\cdot(f, g) = (f'\cdot f, g'\cdot g).$$

To tie all these concepts together we note the following:

6 **OBSERVATION**: The map which assigns to each pair of objects K, K' in the category **K** the set $\mathbf{K}(K, K')$ of morphisms from K to K' becomes a functor

$$\mathrm{hom} : \mathbf{K}^{\mathrm{op}} \times \mathbf{K} \longrightarrow \mathbf{Set} : (K, K') \mapsto \mathbf{K}(K, K')$$

when we make the morphism assignment

$$(K_1 \xleftarrow{\;f\;} K_1', K_2 \xrightarrow{\;g\;} K_2') \mapsto \mathbf{K}(K_1, K_2) \xrightarrow{\;g\cdot(-)\cdot f\;} \mathbf{K}(K_1', K_2').$$

(**hom** is sometimes referred to as the **external representation functor**, and is often written $\mathbf{K}(\cdot, \cdot)$.)

Proof: We must check that hom preserves identities and composition.

(i) The identity of the object (K_1, K_2) of $\mathbf{K}^{\mathrm{op}} \times \mathbf{K}$ is $(K_1 \xleftarrow{\;\mathrm{id}_{K_1}\;} K_1, K_2 \xrightarrow{\;\mathrm{id}_{K_2}\;} K_2)$. hom sends this identity to the morphism $\mathrm{id}_{K_2}\cdot(\,)\cdot\mathrm{id}_{K_1} : \mathbf{K}(K_1, K_2) \longrightarrow \mathbf{K}(K_1, K_2)$, and this is certainly the identity for $\mathbf{K}(K_1, K_2)$, since for any h in $\mathbf{K}(K_1, K_2)$ we have $\mathrm{id}_{K_2}\cdot h\cdot\mathrm{id}_{K_1} = h$. Thus

$$\mathrm{hom}(\mathrm{id}_{(K_1, K_2)}) = \mathrm{id}_{\mathbf{K}(K_1, K_2)}\,.$$

(ii) Again, given $(K_1 \xleftarrow{\;f\;} K_1', K_2 \xrightarrow{\;g\;} K_2')$, and $(K_1' \xleftarrow{\;f'\;} K_1'', K_2' \xrightarrow{\;g'\;} K_2'')$ we have $g'\cdot(g\cdot(h)\cdot f)\cdot f' = (g'\cdot g)\cdot(h)\cdot(f'\cdot f$ in $\mathbf{K}^{\mathrm{op}})$ so that

$$\mathrm{hom}((f', g')\cdot(f, g)) = \mathrm{hom}(f', g')\cdot\mathrm{hom}(f, g)\,. \qquad \square$$

Exercises

1 Let $\mathbf{M} \times \mathbf{M} \longrightarrow \mathbf{M}$ be a partial function satisfying 1 and 2. Introduce the notation $f : k' \longrightarrow k$ to indicate that $D(f) = k'$ and $C(f) = k$. Deduce that this yields a category, with objects the identities, whose 'generalized monoid' is $\mathbf{M} \times \mathbf{M} \longrightarrow \mathbf{M}$.

2 If $H : \mathbf{K} \times \mathbf{L} \longrightarrow \mathbf{N}$ is a functor of two variables and if $K \in \mathbf{K}$ is a fixed object, then show that $H(K, -) : \mathbf{L} \longrightarrow \mathbf{N}$ defined by

$H(K,-)(L) = H(K,L)$, $H(K,-)(f: L \longrightarrow L') = H(\mathrm{id}_K, f)$ is a functor $\mathbf{L} \longrightarrow \mathbf{N}$ (of one variable).

3 Let **K** be the category of *sets with base point*, whose objects are pairs (X,x) with X a set and $x \in X$, and whose morphisms $f: (X,x) \longrightarrow (Y,y)$ are functions f satisfying $f(x) = y$. Show that **K** is a category of sets with structure. Show that $\mathbf{Set}_* \longrightarrow \mathbf{K}$ via $X + \{*\} \leftrightarrow (X + \{*\}, *)$ and that $\mathbf{K} \longrightarrow \mathbf{Set}_*$ via $(X,x) \leftrightarrow (X \backslash \{x\}) + \{*\}$ extend to functors, neither of which is an isomorphism. Is $\mathbf{Set}_*$ a category of sets with structure?

4 Show that if posets are regarded as categories, functors coincide with order-preserving maps. Thus **Poset** is a category of categories and functors!

5 Let $f: (X,d) \longrightarrow (Y,e)$ in **Met**, and let (X,d), (Y,e) be regarded as categories as in exercise 5 of 4.1. Show that $x \leftrightarrow fx$, $\lambda \leftrightarrow \lambda$ makes f a functor.

6 Let **K** be any category and let X_0 in **K** be such that $K \times X_0$ exists for all K in **K**. Prove that $-\times X_0 : \mathbf{K} \longrightarrow \mathbf{K}$ is a functor. Use duality to prove that $-+X_0 : \mathbf{K} \longrightarrow \mathbf{K}$ is a functor if each $K + X_0$ exists.

7 Let **1** be the one-object one-morphism category. Show that objects of any **K** are in bijective correspondence with functors $\mathbf{1} \longrightarrow \mathbf{K}$.

8 Let **2** be the category with two objects D, C and, in addition to id_D and id_C, a single morphism $D \longrightarrow C$. Prove that morphisms in any **K** are in bijective correspondence with functors $\mathbf{2} \longrightarrow \mathbf{K}$.

7.2 FREE AND COFREE

To introduce the functorial approach to 'free' and 'cofree' we recall one of the most familiar definitions of free objects.

1 The **free monoid on the set X of generators** is the set X^* of all finite sequences of elements from X (including the "empty" sequence Λ of length 0) with the associative multiplication of **concatenation**

$$(x_1, \ldots, x_m) \cdot (x_1', \ldots, x_n') = (x_1, \ldots, x_m, x_1', \ldots, x_n')$$

for which Λ is clearly the identity: $\Lambda \cdot w = w = w \cdot \Lambda$ for all $w \in X^*$. Note that an element x yields a string (x) in X^* of length one.

Quite apart from its fundamental role in automata theory (Section 6.3) and formal language theory, $(X^*, \mathrm{conc}, \Lambda)$ is interesting because of the following property:

Given any monoid $M = (S, \circ, e)$ and any map f from X to S viewed as a set, there is a unique homomorphism ψ from $(X^*, \mathrm{conc}, \Lambda)$ to $(S, \circ, e)$ which extends f, i.e. such that $\psi((x)) = f(x)$ for each x in X; $\psi(w \cdot w') = \psi(w) \circ \psi(w')$ for all w, w' in X^*; and $\psi(\Lambda) = e$. In fact, it is clear that the one and only homomorphism

ψ satisfying these requirements is defined by

$$\psi(\Lambda) = e$$

$$\psi((x_1, \ldots, x_n)) = f(x_1) \circ \ldots \circ f(x_n).$$

Using $\eta : X \longrightarrow X^*$ to denote the "inclusion of generators" map $x \mapsto (x)$, we may express the situation in the following categorical form:

2 The monoid $A = (X^*, \text{conc}, \Lambda)$ [for which $UA = X^*$, where $U : \mathbf{Mon} \longrightarrow \mathbf{Set}$ is the forgetful functor] is equipped with a map $\eta : X \longrightarrow UA$ in such a way that, given any other monoid M and any map $f : X \longrightarrow UM$, there exists a unique *homomorphism* ψ such that

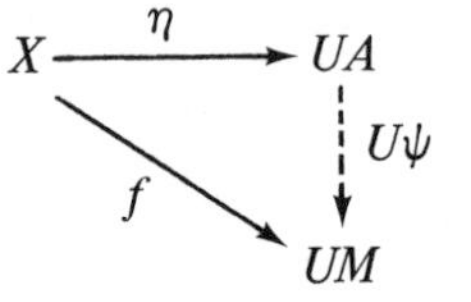

(Maps of Sets)

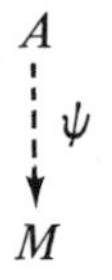

(Homomorphism)

Now, extending our terminology of page 45, we are saying that $X \xrightarrow{\eta} UA$ is couniversal among diagrams of the form $X \xrightarrow{f} UM$, for fixed X, subject to the extra condition that the "bridging morphism" be of the form $U\psi$, where ψ is a **Mon**-morphism, rather than being arbitrary. However, the argument of uniqueness up to isomorphism is still an application of 2.4.6 (why?) and we have:

3 **LEMMA**: A monoid A equipped with a map $\eta : X \longrightarrow UA$ satisfies the condition 2 iff there exists a monoid isomorphism $\psi : A \longrightarrow (X^*, \text{conc}, \Lambda)$ such that $(U\psi \cdot \eta)(x) = (x)$ for all x in X. □

Thus we have once more succeeded in taking an element-by-element definition and showing that it can be captured up to isomorphism in arrow-theoretic terms. This leads to the following general definition:

4 **DEFINITION**: Let $G : \mathbf{A} \longrightarrow \mathbf{B}$ be any functor, and B an object of $\mathbf{B}$. We say the pair (A, η), where A is an object of $\mathbf{A}$ and $\eta : B \longrightarrow GA$ is a morphism of $\mathbf{B}$, is **free over** B **with respect to** G just in case $\eta : B \longrightarrow GA$ has the couniversal property that given any morphism $f : B \longrightarrow GA'$ with A' any object of $\mathbf{A}$, there exists a unique $\mathbf{A}$-morphism $\psi : A \longrightarrow A'$ such that

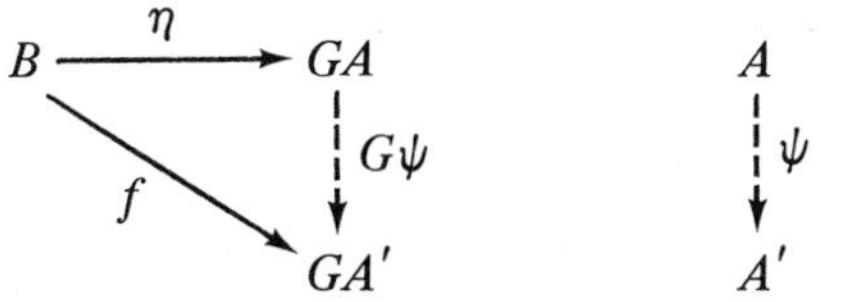 (1)

We refer to η as the **inclusion of generators**; and call the unique ψ satisfying (1) the **A-morphic extension of** f (with respect to G).

As in 3, if (A, η) and (A', η') are both free over B with respect to G, then $G\psi \cdot \eta = \eta'$ for some isomorphism $\psi : A \longrightarrow A'$. However, it may be instructive to give a quite different proof:

Let $(B \downarrow G)$ be the category whose objects are pairs (A', f) for **B**-morphisms $f : B \longrightarrow GA'$ and **A**-objects A'; while the morphisms are **A**-morphisms $\phi : A' \longrightarrow A''$ such that

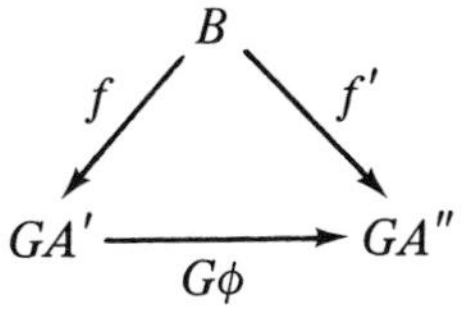

commutes. We immediately recognize that (A, η) is *free* over B iff $B \xrightarrow{\eta} GA$ is *initial* in $(B \downarrow G)$, for to be initial in $(B \downarrow G)$ is simply to be such that given *any* (A', f) there exists a *unique* $\psi : (A, \eta) \longrightarrow (A', f)$, and this is just the situation diagrammed in (1). Uniqueness up to isomorphism of free objects then follows from that of initial objects (2.4.6).

If **C** is a category of sets with structure and G is the forgetful functor $U : \mathbf{C} \longrightarrow \mathbf{Set}$, then (if such an A exists) we refer to A of the (A, η) which is free over B with respect to U as the **free C-object generated by** B.

The reader may check the following (exercises 1 and 2):

5 The **free vector space** generated by B is simply the set of formal sums

$$f = \sum_{b \in B} f_b b \qquad \text{each } f_b \in \mathbf{R} \text{ of } \textit{finite support} \ (\{b \mid f_b \neq 0\} \text{ is finite})$$

with the operations

$$(f + f')_b = f_b + f'_b, \qquad (\lambda \cdot f)_b = \lambda \cdot f_b,$$

i.e. it is the coproduct (2.4.4) of B copies† of **R** itself considered as a vector space.

6 The **free group** generated by B is simply the set of all expressions of the form

† We say "B copies" rather than "$|B|$ copies" since we imagine the family of copies indexed by the elements of B.

$$x_1\, x_2 \ldots x_n \qquad n \geqslant 0$$

where each x_j is of the form either b or $\bar{b}$ for some b in B (we assume $\bar{b} \notin B$ for any $b \in B$) and no $x_j\, x_{j+1}$ is of the form $b\bar{b}$ or $\bar{b}b$ for any b in B. Multiplication is simply concatenation, subject to cancellation of any $b\bar{b}$'s or $\bar{b}b$'s. It is then clear that the identity is the empty string Λ, and that the inverse of $x_1\, x_2 \ldots x_n$ is $\bar{x}_n \ldots \bar{x}_2\, \bar{x}_1$ (with $\bar{\bar{b}}$ replaced by b).

If $B = \{b\}$ with one element, the cancellation rule implies that each element of the free group on one generator is either a string of n b's for some n, the empty string, or a string of n $\bar{b}$'s for some n. If we replace a string of n b's by n, the empty string by 0, and the string of n $\bar{b}$'s by $-n$, we suddenly recognize that the free group on one generator is simply our old friend $(\mathbf{Z}, +, 0)$; and that the free group on an arbitrary set B is the coproduct (3.2.3) of B copies of $(\mathbf{Z}, +, 0)$. In fact, we have the general proposition:

7 **PROPOSITION**: Let $G : \mathbf{K} \longrightarrow \mathbf{Set}$ be a functor. Let 1 be a one-element set and suppose there exists (K, η) free over 1 with respect to G ("(K, η) is the free K-object on one generator"). Let B be a set and suppose that the B-fold coproduct

$$\{K \xrightarrow{\text{in}_b} \bar{K} \mid b \in B\}$$

of B copies of K also exists. Then $(\bar{K}, \bar{\eta})$ is the free K-object over B with respect to G, where $\bar{\eta} : B \longrightarrow G\bar{K}$ is defined by

$$\bar{\eta}(b) = (G(\text{in}_b))(\eta)$$

(where $\eta : 1 \longrightarrow GK$ is identified with an element of GK).

Proof: Let $f : B \longrightarrow GA'$ be given. For each $b \in B, f(b)$ may be thought of as a function from 1 to GA'. For a ψ to complete the diagram

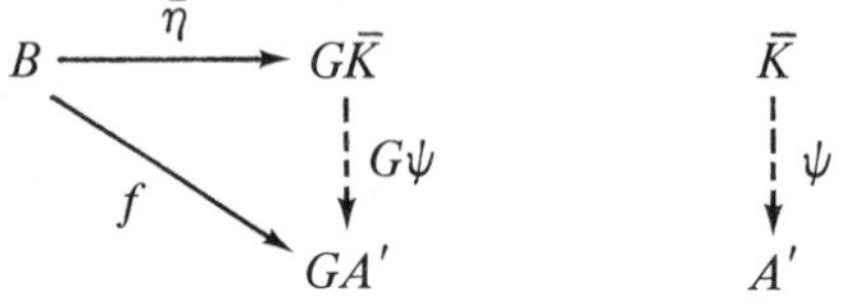

we require that

$$G\psi \cdot \bar{\eta}(b) = f(b) \qquad \text{for all } b \text{ in } B$$

which says that

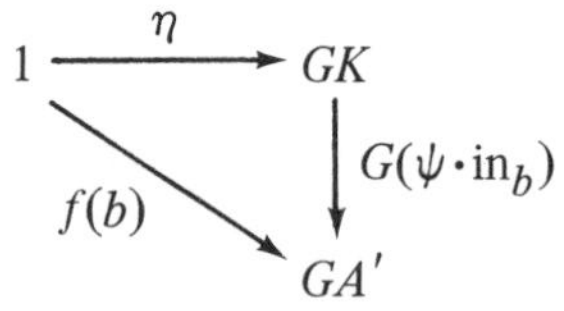

commutes. But since (K, η) is free over 1, the diagram

$$\begin{array}{ccc} 1 & \xrightarrow{\ \eta\ } & GK \\ & \searrow^{f(b)} & \Big\downarrow{\scriptstyle G\psi_b} \\ & & GA' \end{array} \qquad\qquad \begin{array}{c} K \\ \Big\downarrow{\scriptstyle \psi_b} \\ A' \end{array}$$

induces the unique $\psi_b : K \longrightarrow A'$ with $(G\psi_b)\cdot\eta = f(b)$. Since $(\mathrm{in}_b : K \longrightarrow \bar{K})$ is a coproduct diagram there exists a unique $\psi : \bar{K} \longrightarrow A'$ with $\psi \mathrm{in}_b = \psi_b$ for all b:

$$\begin{array}{ccc} \bar{K} & \xleftarrow{\ \mathrm{in}_b\ } & K \\ {\scriptstyle \psi}\Big\downarrow & \swarrow_{\psi_b} & \\ A' & & \end{array}$$

Hence $(\bar{K}, \bar{\eta})$ is indeed free over B with respect to G. □

Dually to 4, we have:

8 DEFINITION: Let $F : \mathbf{A} \longrightarrow \mathbf{B}$ be any functor, and B an object of $\mathbf{B}$. We say the pair (A, ε) where A is an object of $\mathbf{A}$ and $\varepsilon : FA \longrightarrow B$ is a morphism of $\mathbf{B}$, is **cofree over** B **with respect to** F just in case $\varepsilon : FA \longrightarrow B$ has the universal property that given any morphism $f : FA' \longrightarrow B$ with A' an object of $\mathbf{A}$, there exists a unique $\mathbf{A}$-morphism $\psi : A' \longrightarrow A$ such that

$$\begin{array}{ccc} B & \xleftarrow{\ \varepsilon\ } & FA \\ & \nwarrow^{f} & \Big\uparrow{\scriptstyle F\psi} \\ & & FA' \end{array} \qquad\qquad \begin{array}{c} A \\ \Big\uparrow{\scriptstyle \psi} \\ A' \end{array}$$

For example, let F be the functor $-\times X_0 : \mathbf{Set} \longrightarrow \mathbf{Set}$. Then for each B we may take the corresponding (A, ε) to be given by

$$A = B^{X_0}, \qquad \text{the set of all maps from } X_0 \text{ to } B$$

$$\varepsilon : B^{X_0} \times X_0 \longrightarrow B : (f, x) \mapsto f(x), \quad \text{the evaluation map.}$$

Then given $f : A' \times X_0 \longrightarrow B$, the corresponding ψ is clearly

$$\psi : A' \longrightarrow B^{X_0} : a' \mapsto f(a', \cdot)$$

where $f(a', \cdot) : X_0 \longrightarrow B : x \mapsto f(a', x)$. For we then have that

$$\begin{array}{ccc} B & \xleftarrow{\ \varepsilon\ } & B^{X_0} \times X_0 \\ & \nwarrow^{f} & \Big\uparrow{\scriptstyle (a' \mapsto f(a', \cdot);\, x \mapsto x)} \\ & & A' \times X_0 \end{array}$$

certainly commutes, since $\varepsilon(f(a', \cdot), x) = f(a', x)$.

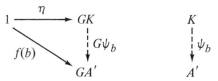

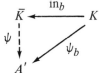

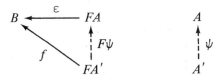

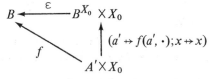

We continue this section by showing that if every B has a free (A, η) with respect to the functor $G : \mathbf{A} \longrightarrow \mathbf{B}$, then we can introduce a functor $F : \mathbf{B} \longrightarrow \mathbf{A}$ for which FB is free over B. Such a functor F is called a **left adjoint** of G. Dually, if every B has a cofree object, then there is a cofree object-constructing functor called the **right adjoint** of G. Thus we say that G **has a left adjoint** just in case every object B has a pair (A, η) free over B; and that F **has a right adjoint** if every A has a pair (B, η) cofree over A.

9 **THEOREM**: Let $G : \mathbf{A} \longrightarrow \mathbf{B}$ be a functor with the property that to every B in $\mathbf{B}$ there corresponds a free object, call it $(FB, \eta B)$† [so that $\eta B : B \longrightarrow GFB$]. Given any $f : B \longrightarrow B'$ in $\mathbf{B}$, define a morphism $Ff : FB \longrightarrow FB'$ by the diagram

$$\begin{array}{ccc} B & \xrightarrow{\eta B} & GFB \\ & \searrow^{\eta B' \cdot f} & \downarrow G(Ff) \\ & & GFB' \end{array} \qquad\qquad \begin{array}{c} FB \\ \downarrow Ff \\ FB' \end{array}$$

Then the collection of maps $F : \mathbf{B} \longrightarrow \mathbf{A}$ so defined is a functor (clearly the object map is unique up to isomorphism, and fixes the morphism maps uniquely) called the **left adjoint** of G.

Proof: From the uniqueness property, it is clear that the commutativity of

$$\begin{array}{ccc} B & \xrightarrow{\eta B} & GFB \\ \downarrow \mathrm{id}_B & & \downarrow \mathrm{id}_{GFB} = G(\mathrm{id}_{FB}) \\ B & \xrightarrow[\eta B]{} & GFB \end{array}$$

implies that $F\mathrm{id}_B = \mathrm{id}_{FB}$. Again, from the commutativity of

$$\begin{array}{ccc} B & \xrightarrow{\eta B} & GFB \\ \downarrow f & & \downarrow GFf \\ B' & \xrightarrow{\eta B'} & GFB' \\ \downarrow g & & \downarrow GFg \\ B'' & \xrightarrow{\eta B''} & GFB'' \end{array} \quad \text{we have} \quad \begin{array}{ccc} B & \xrightarrow{\eta B} & GFB \\ \downarrow g \cdot f & & \downarrow G(Fg \cdot Ff) \\ B'' & \xrightarrow[\eta B'']{} & GFB'' \end{array}$$

(where we have again used the fact that G is a functor), and so uniqueness assures us that $F(g \cdot f)$ must equal $Fg \cdot Ff$. Hence F is a functor. □

† In this discussion, we need to clarify the B with respect to which an η is the "inclusion of generators". In the next section, we shall see that the assignment $B \leftrightarrow \eta B$ is an example of a *natural transformation*.

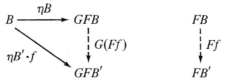

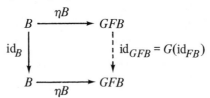

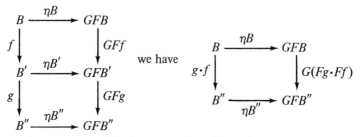

Dually, we have:

10   THEOREM: Let $F : \mathbf{B} \longrightarrow \mathbf{A}$ be a functor with the property that to every A in $\mathbf{A}$ there corresponds a cofree object, call it $(GA, \varepsilon A)$ [so that $\varepsilon A : FGA \longrightarrow A$]. Given any $f : A' \longrightarrow A$ in $\mathbf{A}$, define a morphism $Gf : GA' \longrightarrow GA$ by the diagram

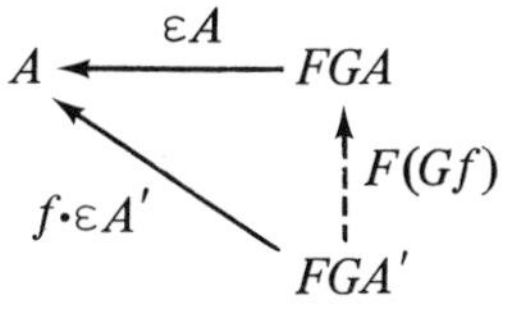

Then $G : \mathbf{A} \longrightarrow \mathbf{B}$ is a functor, which we call the **right adjoint** of F. $\square$

As we shall see in the next section, *F is a left adjoint of G iff G is a right adjoint of F.*

    A functor may have both left and right adjoints, and these will generally be distinct. Returning to our discussion of sequential machines in Section 6.3, the reader may recall that in our discussion of reachability we introduced, just prior to 3.3.6, what we called the free X_0-dynamics on Q generators

11a $\mu_0 Q : (Q \times X_0{}^*) \times X_0 \longrightarrow Q \times X_0{}^* : ((q, w), x) \mapsto (q, wx)$

with which we associated the map

11b $\eta Q : Q \longrightarrow Q \times X_0{}^* : q \mapsto (q, \Lambda)$.

    Again, in our discussion of observability we introduced, just prior to 3.3.11, what we called the cofree X_0-dynamics on Y generators

12a $LY : Y^{X_0{}^*} \times X_0 \longrightarrow Y^{X_0{}^*} : (f, x) \mapsto fL_x$

with which we associated the map

12b $\Lambda Y : Y^{X_0{}^*} \longrightarrow Y : f \mapsto f(\Lambda)$.

    Our task in the remainder of this section is to show that this use of "free" and "cofree" is in fact justified within the usage of the present section. Recall that $\mathbf{Dyn}(X_0)$ is the category whose objects are X_0-dynamics δ or (Q, δ), i.e. maps $\delta : Q \times X_0 \longrightarrow Q$, and whose morphisms are the *dynamorphisms* $h : \delta \longrightarrow \delta'$, i.e. maps $h : Q \longrightarrow Q'$ for which

$$
\begin{array}{ccc}
Q \times X_0 & \xrightarrow{\;\delta\;} & Q \\
{\scriptstyle h \times X_0}\Big\downarrow & & \Big\downarrow{\scriptstyle h} \\
Q' \times X_0 & \xrightarrow{\;\delta'\;} & Q'
\end{array}
$$

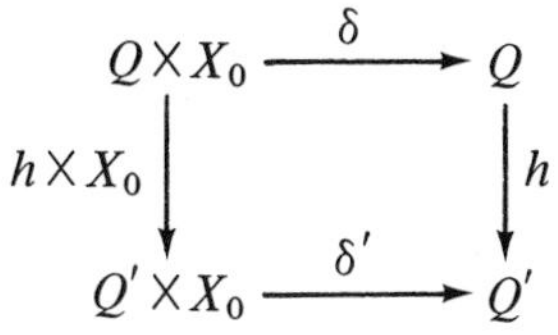

commutes. We may thus define the **forgetful functor**

$$U : \mathbf{Dyn}(X_0) \longrightarrow \mathbf{Set}$$

which sends (Q, δ) to Q and simply regards a dynamorphism as a map at the level of **Set**. We shall see that the free and cofree dynamics are free and cofree with respect to the forgetful functor.

13 **THEOREM**: Let U be the forgetful functor $\mathbf{Dyn}(X_0) \longrightarrow \mathbf{Set}$ and let Q be any set. Then the pair $(\mu_0 Q, \eta Q)$ of 11 is free over Q with respect to U. Thus U has a left adjoint.

Proof: We have to show that, given any X_0-dynamics $\delta' : Q' \times X_0 \longrightarrow Q'$, and any map $f : Q \longrightarrow Q'$, there exists a unique dynamorphism $\psi : \mu_0 Q \longrightarrow \delta'$ such that $f = U\psi \cdot \eta Q$; i.e. a unique map $\psi : Q \times X_0{}^* \longrightarrow Q'$ such that

$$\begin{array}{ccc} Q & \xrightarrow{\eta Q} & Q \times X_0{}^* \\ & {\scriptstyle f} \searrow \quad \text{(A)} & \downarrow {\scriptstyle \psi} \\ & & Q' \end{array} \qquad \text{and} \qquad \begin{array}{ccc} (Q \times X_0{}^*) \times X_0 & \xrightarrow{\mu_0 Q} & Q \times X_0{}^* \\ {\scriptstyle \psi \times X_0} \downarrow & \text{(B)} & \downarrow {\scriptstyle \psi} \\ Q' \times X_0 & \xrightarrow[\delta']{} & Q' \end{array}$$

But (A) just says that

$$\psi(q, \Lambda) = f(q)$$

while (B) says that

$$\psi(q, wx) = \delta'(\psi(q, w), x) \qquad \text{for all } q \text{ in } Q,\ w \text{ in } X_0{}^*,\ x \text{ in } X.$$

It is then immediate from the definition 3.3.5 of the run map $(\delta')^*$ of (Q', δ') that the equations imply

$$\psi(q, w) = (\delta')^*(f(q), w) \qquad \text{for all } q \text{ in } Q,\ w \text{ in } X_0{}^*.$$

Since a unique ψ thus exists, $(\mu_0 Q, \eta Q)$ is indeed free over Q with respect to U. □

14 **THEOREM**: Let U be the forgetful functor $\mathbf{Dyn}(X_0) \longrightarrow \mathbf{Set}$ and let Y be any set. Then the pair $(LY, \Lambda Y)$ of 12 is cofree over Y with respect to U. Thus U has a right adjoint.

Proof: We have to show that, given any X_0-dynamics (Q, δ) and any map $g : Q \longrightarrow Y$, there exists a unique dynamorphism $\phi : \delta \longrightarrow LY$ such that $g = \Lambda Y \cdot U\phi$; i.e. a unique map $\phi : Q \longrightarrow Y^{X_0{}^*}$ such that

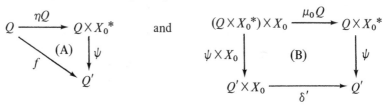

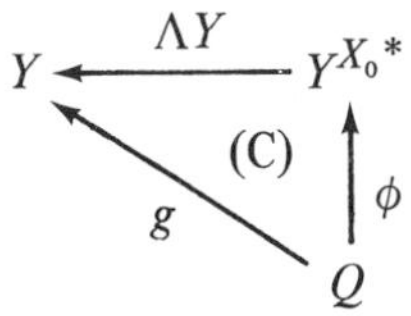

and

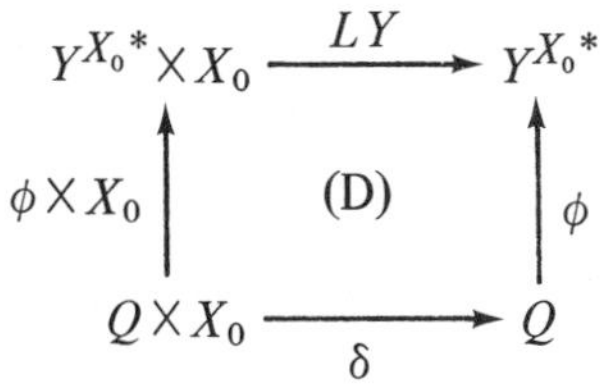

But (C) just says that

$$[\phi(q)](\Lambda) = g(q)$$

while (D) says that

$$[\phi(q)](xw) = [\phi(q)L_x](w) = [\phi(\delta(q, x))](w)$$

for all q in Q, w in $X_0{}^*$ and x in X_0.

These equations imply that

$$[\phi(q)](w) = g[\delta^*(q, w)] \qquad \text{for all } q \text{ in } Q,\ w \text{ in } X_0{}^*$$

(as the reader may readily verify by induction on the length of w). Since a unique ϕ exists, $(LY, \Lambda Y)$ is indeed cofree over Y with respect to U. □

Exercises

1 Verify the free vector space construction of 5.

2 Verify the free group construction of 6.

3 What is the flaw in the following argument?: Dualizing Theorem 7 we immediately have: Let $H : \mathbf{K} \longrightarrow \mathbf{Set}$ where $\mathbf{K}$ is a category with products, and let K_1 be the cofree $\mathbf{K}$-object over the one-element set 1. Then the cofree $\mathbf{K}$-object over B is

$$K = \prod_{b \in B} K_b \qquad \text{where each } K_b \cong K_1$$

and where $\varepsilon : HK \longrightarrow B$ is given by the product diagram

$$HK \xrightarrow{\varepsilon} B \xrightarrow{\pi_b} 1 = HK \xrightarrow{H\pi_b} HK_1 \xrightarrow{\varepsilon 1} 1 .$$ □

4 Let K be a fixed object in $\mathbf{K}^{\text{op}}$ and consider the functor $\mathbf{K}(K, \cdot) : K \longrightarrow \mathbf{Set}$ obtained from the functor $\hom : \mathbf{K}^{\text{op}} \times \mathbf{K} \longrightarrow \mathbf{Set}$ of 1.6 by fixing K as in exercise 2 of 7.1; i.e., $\mathbf{K}(K, -) = \hom(K, -)$. Prove that if $\mathbf{K}$ has coproducts, $\mathbf{K}(K, \cdot)$ has a left adjoint.

5 Let G be a group. An element $x \in G$ *has finite order* if $x^n = e$ for some integer n. G is a **torsion group** if every element of G has finite order. Let **Tor** be the category of torsion groups and group homomorphisms and let $H : \mathbf{Tor} \longrightarrow \mathbf{Grp}$ be the inclusion functor. Show that H has a right adjoint. [Hint: Consider the intersection of all subgroups containing all elements of finite order.]

6 Prove that **Top** $\longrightarrow$ **Set** has a left adjoint and a right adjoint.

7 Let $f: Y \longrightarrow X$ be a function. Let $\mathbf{P}(X)$, $\mathbf{P}(Y)$ be the inclusion-ordered posets of all subsets of X, Y respectively and consider the order-preserving map $f^{-1}: \mathbf{P}(X) \longrightarrow \mathbf{P}(Y)$, $A \mapsto f^{-1}(A) = \{y \in B \mid f(y) \in A\}$. When $\mathbf{P}(X)$, $\mathbf{P}(Y)$ are considered as categories as in 2.2.5, $f^{-1}: \mathbf{P}(X) \longrightarrow \mathbf{P}(Y)$ is a functor. Prove that f^{-1} has both a left adjoint and a right adjoint.

8 Provide examples of contractions $f: (X, d) \longrightarrow (Y, e)$, $g: (Y, e) \longrightarrow (X, d)$ such that, when considered as functors as in exercise 7.1.5, f has g as left adjoint.

9 Let $U: \mathbf{Met} \longrightarrow \mathbf{Set}$ be the underlying set functor. Show that there exists a free metric space over B with respect to U if and only if B has at most one element.

7.3 NATURAL TRANSFORMATIONS AND ADJUNCTIONS

Let us look more closely at the diagram in 2.9 whereby we defined the left adjoint F of the functor G:

$$\begin{array}{ccc} B & \xrightarrow{\eta B} & GFB \\ {\scriptstyle f}\downarrow & & \downarrow{\scriptstyle GFf} \\ B' & \xrightarrow[\eta B']{} & GFB' \end{array}$$

We can rewrite this in the following way (where I is the identity functor $\mathrm{id}_{\mathbf{B}}$)

$$\begin{array}{cccccc} B & & IB & \xrightarrow{\eta B} & (GF)B & \\ {\scriptstyle f}\downarrow & & {\scriptstyle If}\downarrow & & \downarrow{\scriptstyle (GF)f} & \qquad (1) \\ B' & & IB' & \xrightarrow[\eta B']{} & (GF)B' & \end{array}$$

which immediately generalizes to the following (on replacing I by F_1 and GF by F_2):

1 **DEFINITION**: Given two functors F_1 and F_2 from **K** to **L**, a **natural transformation** $\tau: F_1 \dot{\longrightarrow} F_2$ is a function which assigns to each object K of **K** a morphism $\tau K: F_1 K \longrightarrow F_2 K$ in such a way that every **K**-morphism $f: K_1 \longrightarrow K_2$ of **K** yields a commutative diagram:

$$\begin{array}{ccc} K_1 & \qquad & F_1K_1 & \xrightarrow{\tau K_1} & F_2K_1 \\ f\big\downarrow & & F_1f\big\downarrow & & \big\downarrow F_2f \\ K_2 & & F_1K_2 & \xrightarrow[\tau K_2]{} & F_2K_2 \end{array}$$

If we think of each functor F as giving a 'representation' or 'picture' of the category **K** in the category **L**, then the definition says that the representations are naturally related in that we can find a *single* way of transforming the picture F_1K of an object into the picture F_2K of that same object which is consistent with *every* morphism which involves that object as domain or codomain.

2 **EXAMPLE**: Consider the functor $F_1 = - \times X_0 : \mathbf{Set} \longrightarrow \mathbf{Set}$ for X_0 a non-empty set (as discussed following 1.3). Let $F_2 : \mathbf{Set} \longrightarrow \mathbf{Set}$ be the *constant functor induced by* X_0 defined by $F_1K = X_0$ and $F_1(f : K_1 \longrightarrow K_2) = \mathrm{id}_{X_0}$. Then 'projection' is a natural transformation $\tau : F_1 \dot{\longrightarrow} F_2$; specifically, $\tau K : F_1K \longrightarrow F_2K$ is $K \times X_0 \longrightarrow X_0 : (k, x) \mapsto x$.

To check naturality, let $f : K_1 \longrightarrow K_2$ be a function. We have

$$\begin{array}{ccc} K_1 \times X_0 & \xrightarrow{\tau K_1} & X_0 \\ f \times \mathrm{id}_{X_0}\big\downarrow & & \big\downarrow \mathrm{id}_{X_0} \\ K_2 \times X_0 & \xrightarrow[\tau K_2]{} & X_0 \end{array} \qquad \begin{array}{ccc} (k_1, x) & \longmapsto & x \\ \big\downarrow & & \big\downarrow \\ (fk_1, x) & \longmapsto & x \end{array}$$

On the other hand, there exists no natural transformation $\sigma : F_2 \dot{\longrightarrow} F_1$. For consider the consequences of a σ satisfying

$$\begin{array}{ccc} X_0 & \xrightarrow{\sigma K_1} & K_1 \times X_0 \\ \mathrm{id}_{X_0}\big\downarrow & & \big\downarrow f \times \mathrm{id}_{X_0} \\ X_0 & \xrightarrow[\sigma K_2]{} & K_2 \times X_0 \end{array}$$

for each $f : K_1 \longrightarrow K_2$. To see that this is impossible, fix an element $x \in X_0$. Then there are elements $k_1 \in K_1, k_2 \in K_2$ such that $\sigma K_1(x)$ has form (k_1, x') and $\sigma K_2(x)$ has form (k_2, x''). The commutative square above asks for $f(k_1) = k_2$ for all functions f. But, so long as K_2 has more than one element, we can define a function f with $f(k_1) \neq k_2$.

We might observe that a simpler reason why $\sigma : F_2 \dot{\longrightarrow} F_1$ cannot exist is that if K_1 is the empty set there is no function $X_0 \longrightarrow K_1 \times X_0$. However we may easily rework everything in the category **Neset** of non-empty sets and functions. In this case, the map $\tau K : K \times X_0 \to X_0$ has a "right inverse" $\sigma K : X_0 \to K \times X_0$ satisfying $\tau K \cdot \sigma K = \mathrm{id}_{X_0}$ obtained by choosing any element $k \in K$ and defining $\sigma K(x) = (k, x)$. Our point is that the choice of k is not "natural" since σK's cannot be defined to force $\sigma K : F_1 \dot{\to} F_2$.

Notice that if X_0 is a monoid then $F_1 : \mathbf{Mon} \to \mathbf{Mon}$ and $F_2 : \mathbf{Mon} \to \mathbf{Mon}$ can be defined in exactly the same way (cf. exercise 6 of 7.1), only now a $\sigma : F_2 \to F_1$ exists and can be defined by $\sigma K(x) = (e, x)$ where e is the unit of K_1.

We say that $\tau : F_1 \dot{\to} F_2$ is a **natural equivalence** if it has an inverse in the sense that there exists a natural transformation $\tau^{-1} : F_2 \dot{\to} F_1$ such that $\tau^{-1}K \cdot \tau K = \mathrm{id}_{F_1K}$ and $\tau K \cdot \tau^{-1}K = \mathrm{id}_{F_2K}$ for each object K.

For example, our usual feeling that there is no real difference between a set A and its cartesian product $A \times 1$ with a one-element set 1, may be formalized in the observation that the natural transformation

$$\tau : I \dot{\to} - \times 1 : A \mapsto A \times 1$$

(where $I : \mathbf{Set} \to \mathbf{Set}$ is the identity functor) is a natural equivalence with inverse the projection

$$\tau^{-1} : - \times 1 \dot{\to} I : A \times 1 \mapsto A.$$

Clearly, if τ is a natural equivalence then $\tau A : F_1K \to F_2K$ is an isomorphism. The converse is also true: if $\tau : F_1 \dot{\to} F_2$ is a natural transformation and if τK is an isomorphism for all K then τ is a natural equivalence. To prove it, we need only observe that $\tau^{-1} : F_2 \dot{\to} F_1$ defined by $\tau^{-1}K = (\tau K)^{-1}$ is a natural transformation. This amounts to the observation that the commutativity of (A) implies that of (B)

$$\begin{array}{ccccccccc}
K_1 & & F_1K_1 & \xrightarrow{\tau K_1} & F_2K_1 & & F_2K_1 & \xrightarrow{(\tau K_1)^{-1}} & F_1K_1 \\
{\scriptstyle f}\downarrow & & {\scriptstyle F_1 f}\downarrow & \text{(A)} & \downarrow{\scriptstyle F_2 f} & & {\scriptstyle F_2 f}\downarrow & \text{(B)} & \downarrow{\scriptstyle F_1 f} \\
K_2 & & F_1K_2 & \xrightarrow[\tau K_2]{} & F_2K_2 & & F_2K_2 & \xrightarrow[(\tau K_2)^{-1}]{} & F_1K_2
\end{array}$$

as is clear from the computation $(\tau K_2)^{-1} \cdot F_2 f = (\tau K_2)^{-1} \cdot F_2 f \cdot \tau K_1 \cdot (\tau K_1)^{-1}$ $= (\tau K_2)^{-1} \cdot \tau K_2 \cdot F_1 f \cdot (\tau K_1)^{-1} = F_1 f \cdot (\tau K_1)^{-1}$.

We can now use (1) to rewrite Theorem 2.9 in the following more informative way:

3 **THEOREM**: Given any functor $G : \mathbf{A} \to \mathbf{B}$ with the property that to every B there corresponds a free object $(FB, \eta B)$ [so that $\eta B : B \to GFB$] then the object function $F : B \mapsto FB$ extends to a functor (the **left adjoint** of G) in a unique way such that $\eta : I \dot{\to} GF$ is a natural transformation. □

The utility of adjoints is often obtained from the following immediate consequence of their definition:

4 If $G : \mathbf{A} \longrightarrow \mathbf{B}$ has left adjoint F, then there is a one-to-one correspondence

$$\frac{B \xrightarrow{f} GA}{FB \xrightarrow{\psi} A} \tag{2}$$

defined by the diagram

$$\begin{array}{ccc} B & \xrightarrow{\eta B} & GFB \\ & {\scriptstyle f}\searrow & \downarrow{\scriptstyle G\psi} \\ & & GA \end{array}$$

5 If $F : \mathbf{B} \longrightarrow \mathbf{A}$ has a right adjoint G, then there is a one-to-one correspondence

$$\frac{A \xleftarrow{\psi} FB}{GA \xleftarrow{f} B} \tag{3}$$

defined by the diagram

$$\begin{array}{ccc} A & \xleftarrow{\varepsilon A} & FGA \\ & {\scriptstyle \psi}\nwarrow & \uparrow{\scriptstyle Ff} \\ & & FB \end{array}$$

Comparing 4 and 5, it becomes highly plausible that F is a left adjoint of G iff G is a right adjoint of F. We shall indeed see that this is true, but first we tie 3 and 4 together to make explicit the *naturality* of the tableau (2):

We may rephrase the criterion 4 for freeness by saying that the assignment

$$\mathbf{A}(FB, A) \longrightarrow \mathbf{B}(B, GA) : \psi \mapsto G\psi \cdot \eta B \tag{4}$$

yields a bijection. Now consider the diagram

$$\begin{array}{ccccc} A & \qquad & \mathbf{A}(FB, A) & \xrightarrow{\psi \mapsto G\psi \cdot \eta B} & \mathbf{B}(B, GA) \\ {\scriptstyle g}\downarrow & & {\scriptstyle g \cdot (-)}\downarrow & & \downarrow{\scriptstyle Gg \cdot (-)} \\ A' & & \mathbf{A}(FB, A') & \xrightarrow[\psi' \mapsto G\psi' \cdot \eta B]{} & \mathbf{B}(B, GA') \end{array}$$

This commutes, since $G(g \cdot \psi) \cdot \eta B = Gg \cdot G\psi \cdot \eta B$. In other words, for each fixed object B of $\mathbf{B}$, $\phi : \mathbf{A}(FB, -) \dot{\longrightarrow} \mathbf{B}(B, G(-))$ defined by

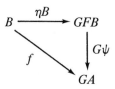

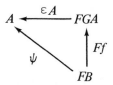

$$\mathbf{A}(FB, A) \xrightarrow{\phi A} \mathbf{B}(B, GA) : \psi \leftrightarrow G\psi \cdot \eta B$$

is a natural equivalence.

Conversely, for fixed B in $\mathbf{B}$, suppose $\phi : \mathbf{A}(FB, -) \longrightarrow \mathbf{B}(B, G(-))$ is a natural equivalence, so that

$$\begin{array}{ccc} A & \qquad & \mathbf{A}(FB, A) & \xrightarrow{\phi A} & \mathbf{B}(B, GA) \\ {\scriptstyle g}\downarrow & & {\scriptstyle g\cdot(-)}\downarrow & & \downarrow{\scriptstyle Gg\cdot(-)} \\ A' & & \mathbf{A}(FB, A') & \xrightarrow[\phi A']{} & \mathbf{B}(B, GA') \end{array}$$

Setting $A = FB$ and defining $\eta B = \phi A(\mathrm{id}_{FB})$ in $\mathbf{B}(B, GFB)$, we have for all g of the form $\psi : FB \longrightarrow A'$ that

$$G\psi \cdot \eta B = G\psi \cdot \phi A(\mathrm{id}_{FB}) = \phi A'(\psi) \tag{5}$$

But since $\phi A'$ is bijective, this says precisely

$$\begin{array}{ccc} B & \xrightarrow{\eta B} & GFB & \qquad & FB \\ & {\scriptstyle f}\searrow & \downarrow{\scriptstyle G\psi} & & \downarrow{\scriptstyle \psi} \\ & & GA' & & A' \end{array}$$

that for all f there exists unique ψ with $G\psi \cdot \eta B = f$, namely $\psi = (\phi A')^{-1}(f)$. We have thus proved:

6 YONEDA PROPOSITION: (A, η) is free over B with respect to $G : \mathbf{A} \longrightarrow \mathbf{B}$ iff the function

$$(\psi : A \longrightarrow A') \leftrightarrow (G\psi \cdot \eta B : B \longrightarrow GA') \tag{6}$$

is a bijection $\mathbf{A}(A, A') \cong \mathbf{B}(B, GA')$ for each A' of $\mathbf{A}$. This bijection is natural in A'. Conversely, given A and B, any isomorphism $\mathbf{A}(A, A') \cong \mathbf{B}(B, GA')$ natural in A' has the form (6) for a unique morphism $\eta : B \longrightarrow GA$ for which (A,η) is free over B. □

At this stage, let us survey the knowledge we have gained in our progression through 2.9, 3 and 4:

(i) Let $G : \mathbf{A} \longrightarrow \mathbf{B}$ be a functor with the property that to every B in $\mathbf{B}$ there corresponds a free object, call it $(FB, \eta B)$. Given any $f : B \longrightarrow B'$ in $\mathbf{B}$, define a morphism $Ff : FB \longrightarrow FB'$ by the diagram

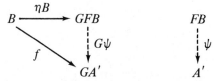

$$\begin{array}{ccc} B & \xrightarrow{\eta B} & GFB \\ {\scriptstyle f}\downarrow & & \vdots\,{\scriptstyle G(Ff)} \\ B' & \xrightarrow[\eta B']{} & GFB' \end{array} \qquad (7)$$

Then the $F:\mathbf{B}\longrightarrow\mathbf{A}$ so defined is a functor, called the *left adjoint* of G.

(ii) The diagram (7) then shows that $\eta : I \dot{\longrightarrow} GF$ is a natural transformation.

(iii) The "freeness diagram"

$$\begin{array}{ccc} B & \xrightarrow{\eta B} & GFB \\ & {\scriptstyle f}\searrow & \downarrow{\scriptstyle G\psi} \\ & & GA \end{array} \qquad\qquad \begin{array}{c} FB \\ \downarrow{\scriptstyle \psi} \\ A \end{array}$$

establishes a bijection $(FB \xrightarrow{\psi} A) \leftrightarrow (B \xrightarrow{\phi(\psi)} GA)$, where $\phi(\psi) = G\psi\cdot\eta B$, which for each fixed B is natural in A.

This suggests the following definition which is reminiscent of the definition† $\langle G^*b, a\rangle = \langle b, Ga\rangle$ of adjoint maps between inner product spaces:

7 **DEFINITION**: Let $\mathbf{A}$ and $\mathbf{B}$ be categories. An **adjunction** from $\mathbf{B}$ to $\mathbf{A}$ is a triple $\langle F, G, \phi\rangle : \mathbf{B} \rightharpoonup \mathbf{A}$, where F and G are functors

$$\mathbf{B} \underset{G}{\overset{F}{\rightleftarrows}} \mathbf{A}$$

while ϕ is a function which assigns to each pair of objects $B \in \mathbf{B}$ and $A \in \mathbf{A}$ a bijection

$$\phi = \phi(B, A) : \mathbf{A}(FB, A) \cong \mathbf{B}(B, GA)$$

which is natural in B and A.

"Naturality in B" is clarified in the proof of the expected:

8 **THEOREM**: If $G:\mathbf{A}\longrightarrow\mathbf{B}$ has left adjoint $F:\mathbf{B}\longrightarrow A$ and inclusion of generators $\eta : I \dot{\longrightarrow} GF$, then the bijection

$$\phi(B, A) : \mathbf{A}(FB, A) \cong \mathbf{B}(B, GA) : \psi \leftrightarrow G\psi\cdot\eta B$$

is such that $\langle F, G, \phi\rangle : \mathbf{B} \rightharpoonup \mathbf{A}$ is an adjunction.

Proof: We have only to show naturality in B, i.e. that

† The reader unfamiliar with this definition may safely ignore this comment.

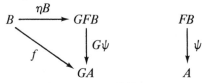

$$\begin{array}{ccccc} B' & \qquad & \mathbf{A}(FB', A) & \xrightarrow{\phi(B', A)} & \mathbf{B}(B', GA) \\ h \Big\uparrow & & (-)\cdot Fh \Big\downarrow & & \Big\downarrow (-)\cdot h \\ B & & \mathbf{A}(FB, A) & \xrightarrow[\phi(B, A)]{} & \mathbf{B}(B, GA) \end{array}$$

commutes for all $h : B \longrightarrow B'$ in $\mathbf{B}$. But for $\psi : FB' \longrightarrow A$ in $\mathbf{A}$ we have

$$\begin{aligned} \phi(B, A)(\psi \cdot Fh) &= G\psi \cdot GFh \cdot \eta B \\ &= G\psi \cdot \eta B' \cdot h \qquad (\eta \text{ is a natural transformation}) \\ &= [\phi(B', A)(\psi)] \cdot h\,. \end{aligned}$$ □

We now characterize, in 9 – 12, the relationship between the natural equivalence $\phi : \mathbf{A}(FB, A) \dot{\longrightarrow} \mathbf{B}(B, GA)$ and the natural transformations $\eta : I_{\mathbf{B}} \dot{\longrightarrow} GF$ and $\varepsilon : FG \longrightarrow I_{\mathbf{A}}$.

9 **PROPOSITION**: Let $\mathbf{A}$, $\mathbf{B}$ be categories and let $F : \mathbf{B} \longrightarrow \mathbf{A}$, $G: \mathbf{A} \longrightarrow \mathbf{B}$ be functors. Then each ϕ such that $\langle F, G, \phi \rangle$ is an adjunction from $\mathbf{B}$ to $\mathbf{A}$ induces

(i) A natural transformation $\eta : I_{\mathbf{B}} \dot{\longrightarrow} GF$ with respect to which G has F as left adjoint with inclusion of generators η, and such that the unique $\psi : FB \longrightarrow A$ such that

$$G\psi \cdot \eta B = f \quad \text{is given by} \quad \psi = (\phi(B, A))^{-1}(f) \tag{8}$$

(ii) A natural transformation $\varepsilon : FG \dot{\longrightarrow} \mathrm{id}_{\mathbf{A}}$ with respect to which F has G as a right adjoint with evaluation ε, and such that the unique $f : B \longrightarrow GA$ such that

$$\varepsilon A \cdot Ff = \psi \quad \text{is given by} \quad f = \phi(B, A)(\psi) \tag{9}$$

Moreover

(iii) We have

$$\begin{aligned} G \xrightarrow{\eta G} GFG \xrightarrow{G\varepsilon} G &= \mathrm{id}_G \\ F \xrightarrow{F\eta} FGF \xrightarrow{\varepsilon F} F &= \mathrm{id}_F \end{aligned} \tag{10}$$

(where ηG is the natural transformation defined by $(\eta G)A = = \eta(GA) : GFGA \longrightarrow GA$, $(G\varepsilon)A = G(\varepsilon A)$, etc.).

Proof: The pleasant thing about the definition 7 of adjunction is that it is self-dual, i.e. $\langle G, F, \phi^{-1} \rangle$ is an adjunction from $\mathbf{A}^{\mathrm{op}}$ to $\mathbf{B}^{\mathrm{op}}$. Thus (ii) will follow by duality, as soon as we have proved (i). Let, then, $\langle F, G, \phi \rangle : \mathbf{B} \longrightarrow \mathbf{A}$ be an adjunction, so that the bijection $\phi : \mathbf{A}(FB, A) \cong \mathbf{B}(B, GA)$ is, in particular, natural in A for each B. Thus, applying the Yoneda Proposition 6 we have that there is a unique morphism $\eta : B \longrightarrow GFB$ such that $\phi(\psi) = Gf \cdot \eta B$, and such that $(FB, \eta B)$ is free over B with respect to G. This establishes (i); and, as we have said, (ii) follows by duality. For (iii), take $f = \mathrm{id}_A$ in (9), to deduce that

$$\varepsilon A = \phi^{-1}(GA, A)$$

Inserting this in (8) we deduce that

$$\mathrm{id}_{GA} = \phi(\varepsilon A) = G(\varepsilon A) \cdot \eta GA$$

so that the natural transformation

$$G \xrightarrow{\eta G} GFG \xrightarrow{G\varepsilon} G$$

is the identity transformation $G \dashrightarrow G$. Dually, $F \xrightarrow{F\eta} FGF \xrightarrow{\varepsilon F} F$ is the identity transformation $F \dashrightarrow F$. □

10 PROPOSITION: Let $F : \mathbf{B} \longrightarrow \mathbf{A}$, $G : \mathbf{A} \longrightarrow \mathbf{B}$ be functors and let $\eta : I_{\mathbf{B}} \longrightarrow GF$, $\varepsilon : GF \longrightarrow I_{\mathbf{A}}$ be natural transformations such that (10) holds. Then $\langle F, G, \phi \rangle$ is an adjunction from $\mathbf{B}$ to $\mathbf{A}$ where $\phi(B, A)$ is the passage of (4).

Proof: As was seen following (4), naturality in A holds because G is a functor. Naturality in B follows from the fact that η is a natural transformation, as is clear from the proof of 8. Define $\phi^{-1}(B, A)$ by

$$\phi^{-1}(B, A) : \mathbf{B}(B, GA) \longrightarrow \mathbf{A}(FB, A)$$
$$f \mapsto \varepsilon A \cdot Ff$$

Then
$$\begin{aligned} \phi(B, A) \cdot \phi^{-1}(B, A)(f) &= \phi(B, A)(\varepsilon A \cdot Ff) \\ &= G\varepsilon A \cdot GFf \cdot \eta B \\ &= G\varepsilon A \cdot \eta GA \cdot f \qquad (\eta \text{ is a natural transformation}) \\ &= f \qquad (\text{by } (10)) \end{aligned}$$

and, similarly, $\phi^{-1}\phi = \mathrm{id}$ using the naturality of ε and the other equation of (10). □

11 PROPOSITION: Given functors $F : \mathbf{B} \longrightarrow \mathbf{A}$, $G : \mathbf{A} \longrightarrow \mathbf{B}$ the class of adjunctions $\langle F, G, \phi \rangle$ is in bijective correspondence with the class of quadruples $\langle F, G, \eta, \varepsilon \rangle$ such that $\eta : I_{\mathbf{B}} \dashrightarrow GF$, $\varepsilon : FG \dashrightarrow I_{\mathbf{A}}$ are natural transformations subject to (10) by the mutually inverse passages described in 9 and 10.

Proof: Starting with ϕ, passing to (η, ε) as in 9 and thence to $\bar{\phi}$ by 10 we have for $\psi : FB \longrightarrow A$,

$\bar{\phi}(B, A)(\psi)$

$= G\psi \cdot \eta B$

$= G\psi \cdot [\phi(B, FB)(\mathrm{id}_{FB})]$ (recall how ηB was defined just preceding (5))

$= \phi(B, A)(\psi \cdot \mathrm{id}_{FB})$ (by naturality of ϕ in its second variable)

$= \phi(B, A)(\psi)$.

Conversely, starting with (η, ε), passing to ϕ and thence to $(\bar{\eta}, \bar{\varepsilon})$ we have

$$\begin{aligned}\bar{\eta}B &= \phi(B, FB)(\mathrm{id}_{FB})\\ &= G(\mathrm{id}_{FB})\cdot\eta B \quad \text{(by (4))}\\ &= \eta B\end{aligned}$$

and $\bar{\varepsilon} = \varepsilon$ similarly. □

We sum up the salient points of 9 – 11 with the

12 THEOREM: Let $F : \mathbf{B} \longrightarrow \mathbf{A}$, $G : \mathbf{A} \longrightarrow \mathbf{B}$ be functors. Then the following four conditions are equivalent:

(i) F has G as a left adjoint.

(ii) G has F as a right adjoint.

(iii) There exists an equivalence $\phi(B, A) : \mathbf{A}(FB, A) \longrightarrow \mathbf{B}(B, GA)$ natural in A and B, i.e. there exists ϕ such that $\langle F, G, \phi\rangle$ is an adjunction from $\mathbf{B}$ to $\mathbf{A}$.

(iv) There exist natural transformations $\eta : I_{\mathbf{B}} \dot{\longrightarrow} GF$ and $\varepsilon : FG \longrightarrow I_{\mathbf{A}}$ such that $G\varepsilon A \cdot \eta GA = \mathrm{id}_{GA}$ for all A in $\mathbf{A}$ and $\varepsilon FB \cdot F\eta B = \mathrm{id}_{FB}$ for all B in $\mathbf{B}$.

Moreover, the (η, ε) of (iv) coincide with the 'inclusion of the generators' of (i) and the 'evaluation' of (ii). □

We close this section by noting that the proof of the "Yoneda Proposition" 6 rested on the fact that the natural transformation $\phi : \mathbf{A}(A, -) \longrightarrow \mathbf{B}(B, G-)$ was completely determined by $\phi A(\mathrm{id}_A)$, as we see from equation (5). The reader who pursues her study of category theory will find such an observation of such general applicability that it is worth explicit statement in general form:

13 THE YONEDA LEMMA: Given any category $\mathbf{A}$, any functor $H : \mathbf{A} \longrightarrow \mathbf{Set}$, and any object A of $\mathbf{A}$, there is a bijection

$$\mathrm{Nat}(\mathbf{A}(A, -), H) \cong HA : \tau \leftrightarrow \tau A(\mathrm{id}_A)$$

[where, given any two functors $F_1, F_2 : \mathbf{K} \longrightarrow \mathbf{L}$, $\mathrm{Nat}(F_1, F_2)$ denotes the *set* of all natural transformation $F_1 \dot{\longrightarrow} F_2$].

Proof: Just as we obtained (5), we may deduce that, given any A', and any $g \in \mathbf{A}(A, A')$, we have from

$$\begin{array}{ccc} \mathbf{A}(A, A) & \xrightarrow{\tau A} & HA \\ {\scriptstyle g\cdot(-)}\Big\downarrow & & \Big\downarrow{\scriptstyle Hg} \\ \mathbf{A}(A, A') & \xrightarrow[\tau A']{} & HA' \end{array}$$

$$\tau A'(g) = Hg \cdot \tau A(\mathrm{id}_A). \tag{11}$$

Conversely, given an element a of HA, the equation

$$\tau A'(g) = Hg(a)$$

certainly defines a natural transformation $\tau : \mathbf{A}(A, -) \dot{\longrightarrow} H$, since

$$\begin{array}{c} A' \\ {\scriptstyle h}\Big\downarrow \\ A'' \end{array} \qquad\qquad \begin{array}{ccc} \mathbf{A}(A, A') & \xrightarrow{\tau A'} & HA' \\ {\scriptstyle h\cdot(-)}\Big\downarrow & & \Big\downarrow{\scriptstyle Hh} \\ \mathbf{A}(A, A'') & \xrightarrow[\tau A'']{} & HA'' \end{array}$$

commutes just in case

$$Hh \cdot Hg(a) = H(h \cdot g)(a),$$

and this is immediate since H is a functor. The reader can easily check that these passages are inverse to each other. □

Exercises

1 Let $F : \mathbf{B} \longrightarrow \mathbf{A}$, $G : \mathbf{A} \longrightarrow \mathbf{B}$ be functors. Show that $\mathbf{A}(F-, -)$ and $\mathbf{B}(-, G-)$ are functors from $\mathbf{B}^{\mathrm{op}} \times \mathbf{A}$ to **Set**. Show that $\phi(B, A) : \mathbf{A}(FB, A) \longrightarrow \mathbf{B}(B, GA)$ is natural in A and in B if and only if it is a natural transformation.

2 Let $U : \mathbf{Mon} \longrightarrow \mathbf{Set}$ be the forgetful functor and let $U^2 : \mathbf{Mon} \longrightarrow \mathbf{Set}$ be the functor $(X, m, e) \mapsto X \times X$, $f \mapsto f \times f$. Show that $\tau(X, m, e) = m$ is a natural transformation $U^2 \dot{\longrightarrow} U$. Similarly, show that $\bar{\tau}(X, m, e) = \bar{m}$ where $\bar{m}(x, y) = m(y, x)$ is also a natural transformation. Further, show that if $\sigma : U^2 \dot{\longrightarrow} U$ is a natural transformation then σ is associative (i.e. the binary operation $\sigma(X, m, e) : X^2 \longrightarrow X$ satisfies the associative law for all (X, m, e)) if and only if σ is a constant, a projection, τ, or $\bar{\tau}$. [Hint: Use the Yoneda Lemma to prove that natural transformations correspond to elements of the free monoid on two generators.]

3 In the context of the forgetful functor $G : \mathbf{Mon} \longrightarrow \mathbf{Set}$, show that corresponding adjunction data are given as indicated: $FB = (B^*, \mathrm{conc}, \Lambda)$, for $f : B \longrightarrow B'$, $(Ff)(b_1 \ldots b_n) = (fb_1 \ldots fb_n)$, $(\eta B)(b) = (b)$,

$\varepsilon(S, \circ, e) : (S^*, \text{conc}, \Lambda) \longrightarrow S = s_1 \ldots s_n \mapsto s_1 \circ \ldots \circ s_n$.

4 With respect to $G : \mathbf{Vect} \longrightarrow \mathbf{Set}$ show that $\varepsilon(V, +, \cdot)(\Sigma f_b b) = \Sigma f_b \cdot b$ (where the latter sum is the iterated + in V).

5 In the context of the example following 2.8 show that $GA = A^{X_0}$, $Gf : A^{X_0} \longrightarrow (A')^{X_0}$ sends $g : X_0 \longrightarrow A$ to $f \cdot g : X_0 \longrightarrow A'$, and that $\eta B : B \longrightarrow (B \times X_0)^{X_0}$ sends b to the map $X_0 \longrightarrow B \times X_0$, $x \mapsto (b, x)$.

6 In the context of 2.13 show that $\varepsilon(Q, \delta)$ is δ^*.

7 In the context of 2.14, show that $\eta(Q, \delta) : (Q, \delta) \longrightarrow Q^{X_0^*}$ is the map $q \mapsto \delta^*(q, -)$.

8 Let $U : \mathbf{B} \longrightarrow \mathbf{A}$ have a left adjoint $L : \mathbf{A} \longrightarrow \mathbf{B}$ and a right adjoint $R : \mathbf{A} \longrightarrow \mathbf{B}$. Show that $UL : \mathbf{A} \longrightarrow \mathbf{A}$ has UR as a right adjoint. [Hint: Consider $\mathbf{A}(ULA, A') \cong \mathbf{B}(LA, RA') \cong \mathbf{A}(A, URA')$]. Solve for η, ε, ϕ in terms of these parameters for the given adjunctions. What happens when $U = \text{Dyn}(X_0) \longrightarrow \mathbf{Set}$ in the contexts of 2.13 and 2.14?

9 Prove the **Composition Theorem**: If $F_1 : \mathbf{C} \longrightarrow \mathbf{B}$ and $F_2 : \mathbf{B} \longrightarrow \mathbf{A}$ both have left adjoints G_1 and G_2, respectively, then $F_2 F_1 : \mathbf{C} \longrightarrow \mathbf{A}$ also has a left adjoint. What is it?

Chapter 8

THE ADJOINT FUNCTOR THEOREM

We have said nothing so far about practical ways to detect the existence of free objects. Section 2 will present a standard characterization theorem, but we first introduce some necessary conditions for a functor to have a left adjoint.

8.1 NECESSARY CONDITIONS

In this section, we show that any functor with a left adjoint preserves both products and equalizers in the following sense:

1 **DEFINITION**: Let $G : \mathbf{A} \longrightarrow \mathbf{B}$ be a functor. G **preserves products** if whenever $(P_i : A \longrightarrow A_i)$ is a product in **A**, then $(GP_i : GA \longrightarrow GA_i)$ is a product in **B**.

Similarly, G **preserves equalizers** if whenever $i = \text{eq}(f, g)$ in **A**

$$E \xrightarrow{\;i\;} A_1 \underset{g}{\overset{f}{\rightrightarrows}} A_2$$

then $Gi = \text{eq}(Gf, Gg)$ in **B**.

Dually, we may define preservation of coproducts and coequalizers.

2 **THEOREM**: Let $G : \mathbf{A} \longrightarrow \mathbf{B}$ be a functor with left adjoint $F : \mathbf{B} \longrightarrow \mathbf{A}$. Then G preserves products.

Proof: Let $P_i : A \longrightarrow A_i$ be a product diagram in **A** and let $f_i : B \longrightarrow GA_i$ be a similarly-indexed family of morphisms

$$\begin{array}{ccc} B & & GA \\ & \searrow^{f_i} & \downarrow GP_i \\ & & GA_i \end{array}$$

in **B**. Let (FB, η) be free over B with respect to G. For each i, there exists unique $\psi_i : FB \longrightarrow A_i$ satisfying $G\psi_i \cdot \eta = f_i$.

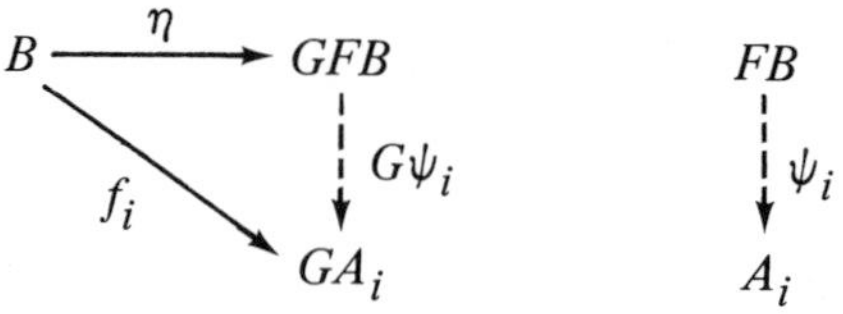

By the product property, there exists a unique map

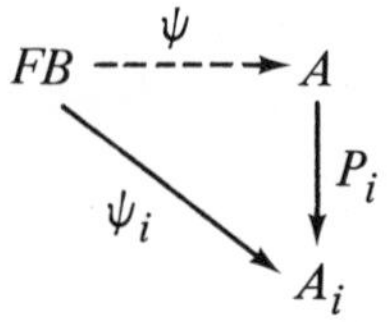

$\psi : FB \longrightarrow A$ such that $P_i \cdot \psi = \psi_i$ for all i.

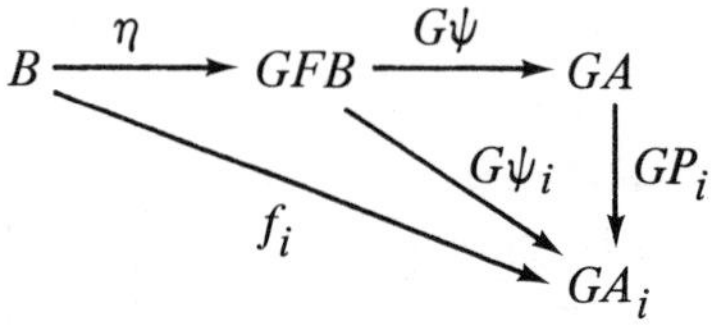

Define $f : B \longrightarrow GA = G\psi \cdot \eta$. Then for all i,

$$GP_i \cdot f = GP_i \cdot G\psi \cdot \eta = G(P_i \cdot \psi) \cdot \eta = G\psi_i \cdot \eta = f_i .$$

Suppose that $g : B \longrightarrow GA$ also satisfied $GP_i \cdot g = f_i$ for all i.

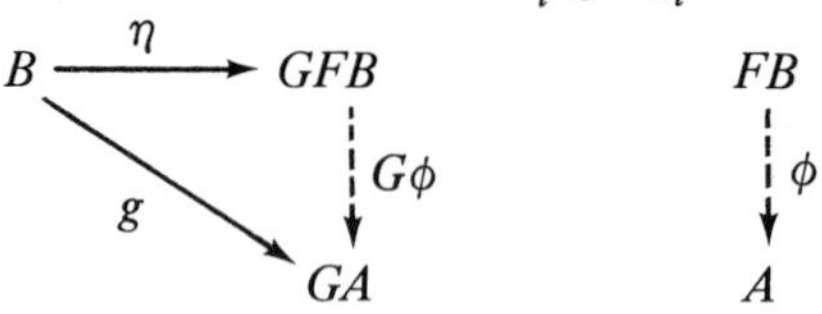

Then there exists unique $\phi : FB \longrightarrow A$ with $G\phi \cdot \eta = g$. For each i we have

$$G(P_i \cdot \phi) \cdot \eta = GP_i \cdot G\phi \cdot \eta = GP_i \cdot g = f_i = G\psi_i \cdot \eta .$$

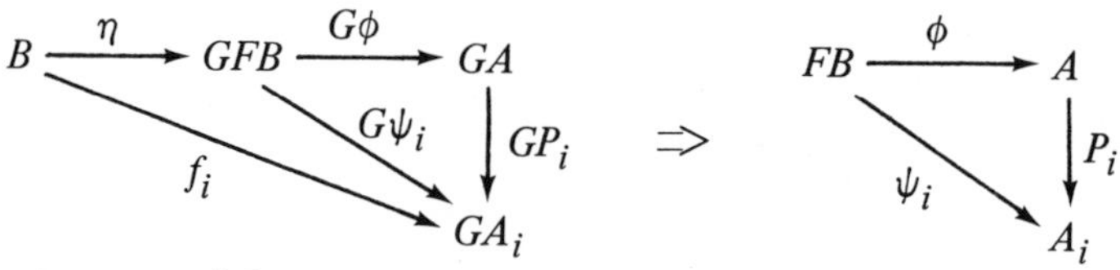

By the uniqueness of the **A**-morphic extension of f_i, $P_i \phi = \psi_i$. By the uniqueness in the universal property defining 'product', $\phi = \psi$. Therefore $g = G\phi \cdot \eta = G\psi \cdot \eta = f$ as desired. This completes the proof that G preserves products. The reader should verify that we have not neglected to prove that G preserves terminal objects (i.e., $G1$ is terminal in **B** if 1 is terminal in **A**). The above proof – with I empty – is valid. □

3 **THEOREM**: Let $G : \mathbf{A} \longrightarrow \mathbf{B}$ be a functor with left adjoint $F : \mathbf{B} \longrightarrow \mathbf{A}$. Then G preserves equalizers.

Proof: Let $i = \text{eq}(f, g)$ in **A**, with

$$E \xrightarrow{i} A_1 \underset{g}{\overset{f}{\rightrightarrows}} A_2,$$

and let $h : B \longrightarrow GA_1$ in **B** satisfy $Gf{\cdot}h = Gg{\cdot}h$.

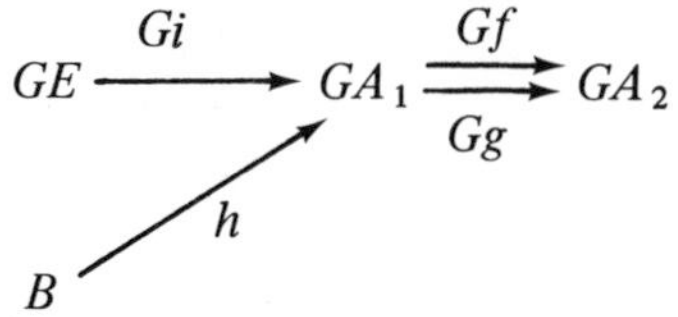

Let (FB, η) be free over B with respect to G and let $\psi : FB \longrightarrow A_1$ be the unique **A**-morphic extension of h.

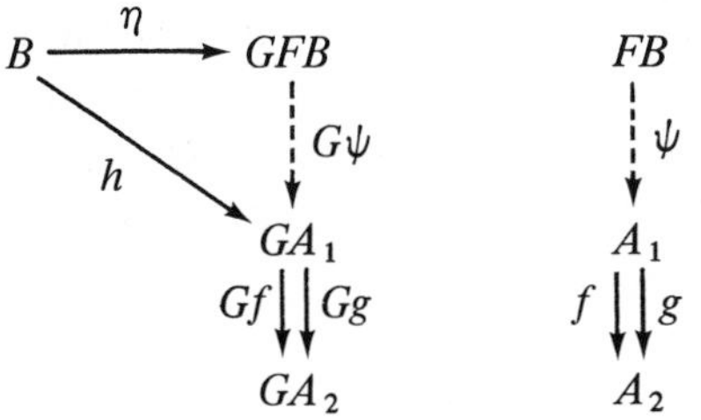

As $G(f\psi){\cdot}\eta = Gf{\cdot}G\psi{\cdot}\eta = Gf{\cdot}h = Gg{\cdot}h = G(g\psi){\cdot}\eta$ we have $f\psi = g\psi$. Thus there exists unique $\phi : FB \longrightarrow E$

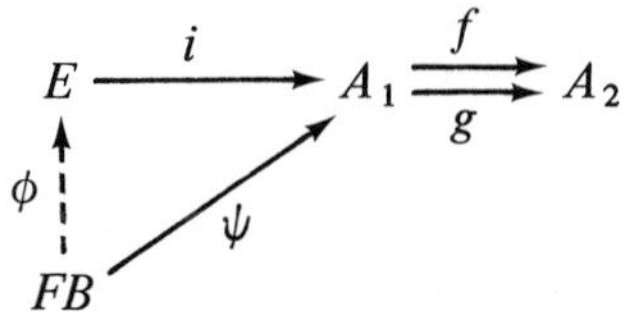

with $i{\cdot}\phi = \psi$. The commutative diagram

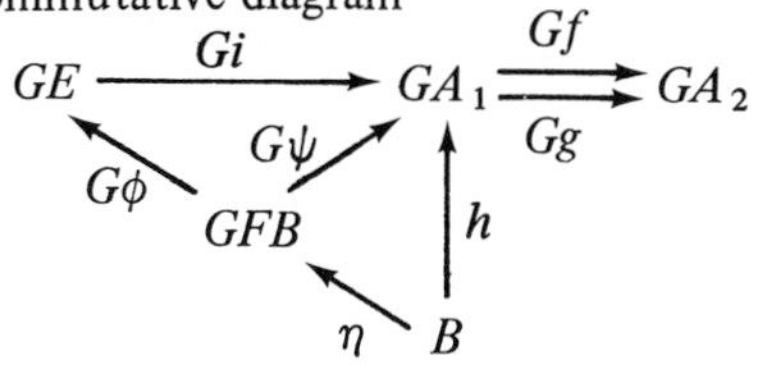

leads us to define $\alpha : B \longrightarrow GE = G\phi{\cdot}\eta$. Then $Gi{\cdot}\alpha = Gi{\cdot}G\phi{\cdot}\eta = G(i\phi){\cdot}\eta =$ $= G\psi{\cdot}\eta = h$. The proof that if $\bar{\alpha} : B \longrightarrow GE$ satisfies $Gi{\cdot}\bar{\alpha} = h$ then $\alpha = \bar{\alpha}$ is similar to that part of the proof in "G preserves products" and will be left as exercise 4 for the reader. □

Duality immediately yields:

4 **COROLLARY:** Any functor which has a right adjoint preserves coproducts and coequalizers. □

Exercises

1 Show that the functor $\mathbf{K}(K, -) : \mathbf{K} \longrightarrow \mathbf{Set}$ of exercise 7.2.4 preserves products and equalizers.

2 Show that the underlying set functor $U : \mathbf{Met} \longrightarrow \mathbf{Set}$ preserves products and equalizers. [Hint: U is essentially $\mathbf{Met}(1, -)$ where 1 is the one-element (terminal) object of **Met**; now use exercise 1.] Prove, however, that U does not have a left adjoint. [Hint: Combine Proposition 7.2.7 with 4.2.4.]

3 Let I be a set and let $\mathbf{K}$ be a category. Define the category $\mathbf{K}^I$ to have as objects all I-tuples $(K_i \mid i \in I)$ of $\mathbf{K}$-objects and as morphisms $f : (K_i) \longrightarrow (L_i)$ all I-tuples $f = (f_i)$ with $f_i : K_i \longrightarrow L_i$, with composition $(f \cdot g)_i = f_i \cdot g_i$ and identities $(\mathrm{id})_{(K_i)} = (\mathrm{id}_{K_i})$. Show that the functor $E : \mathbf{K} \longrightarrow \mathbf{K}^I$ defined by $(EK)_i = K$, $(Ef : K \longrightarrow L)_i = f : K \longrightarrow L$ preserves products and equalizers. Prove that a free E-object over (K_i) generators is the same concept as $\amalg K_i$. Thus E does not always have a left adjoint.

4 Complete the proof of Theorem 3.

8.2 SUFFICIENT CONDITIONS

We have just seen that every functor with a left adjoint must preserve products and equalizers. In this section we introduce one more condition which ensures that a functor which preserves products and equalizers will have a left adjoint.

1 **DEFINITION:** Let $G : \mathbf{A} \longrightarrow \mathbf{B}$ be a functor and let B be an object of $\mathbf{B}$. A **solution set for** B is a set $\mathbf{S}$ of pairs (S, s), where each S is an object of $\mathbf{A}$ and s is a morphism in $\mathbf{B}$ of the form $s : B \longrightarrow GS$, with the property that given an arbitrary $\mathbf{B}$-morphism of form $f : B \longrightarrow GA$ there exists $(S, s) \in \mathbf{S}$ and $\psi : S \longrightarrow A$ in A such that

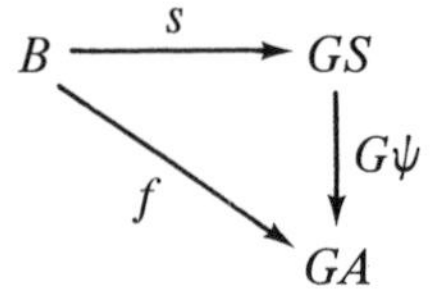

$G\psi \cdot s = f$. G **satisfies the solution set condition** if for all objects B there exists a solution set for B.

It is clear that if G has a left adjoint F then $\mathbf{S} = \{(FB, \eta B)\}$ is a one-element solution set for B and G satisfies the solution set condition.

The astute reader has perhaps already wondered why an arbitrary functor does not satisfy the solution set condition. For why not define **S** to be the 'set' of all pairs (A, f) with $f : B \longrightarrow GA$? The reason this will not suffice is that the 'set' in question is too 'large', just as in exercise 2.4.13, where we saw that the product P of all non-empty sets in **Set** is too 'large' to itself be in **Set**. Rather than trying to advance a precise explanation let us offer the maxim that, in the definition of solution set, **S** must be a 'small set'. We shall refine our usage of exercise 2.4.13 by requiring 'small sets' to belong to a *suitable* collection of sets – but we defer our discussion on what is "suitable" until after the proof of the following theorem. The reader is advised to read the proof with an alertness for the size of the sets actually used.

We may paraphrase the adjoint functor theorem, which we now prove, by saying that if G preserves products and equalizers, we may 'splice together' the pairs (S, s) of a solution set for an object B to construct a pair (FB, η) which is free over B:

2 **ADJOINT FUNCTOR THEOREM**: Let **A** be a category such that every set of objects in **A** has a product and such that every pair $f, g : A_1 \longrightarrow A_2$ in **A** has an equalizer. Let $G : \mathbf{A} \longrightarrow \mathbf{B}$ be a functor. Then G has a left adjoint if and only if the following two conditions hold:

1. G preserves products and G preserves equalizers.
2. G satisfies the solution set condition.

Proof: We have already seen the necessity of these conditions; it only remains to check their sufficiency: Fix an object B of **B**. We must construct (FB, η) free over B with respect to G. Let **S** be a solution set for B. Let $A = \Pi(S \mid (S, s) \in \mathbf{S})$. Define $\Gamma : B \longrightarrow GA$ by $G\pi_{(S, s)} \cdot \Gamma = s$ for all $(S, s) \in \mathbf{S}$, i.e. by the diagram

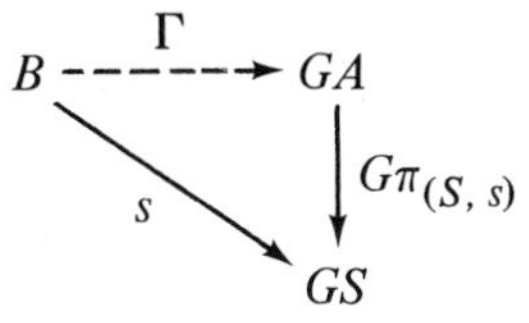

which is a product diagram because G preserves products. Thus Γ is uniquely defined.

We observe that (A, Γ) is a one-element solution set for B, that is, (A, Γ) satisfies the existence (though perhaps not the uniqueness) criterion in the definition of "free object over B with respect to G". To check existence, let $f : B \longrightarrow GA'$. There exists $(S, s) \in \mathbf{S}$ and $\phi : S \longrightarrow A'$ with $G\phi \cdot s = f$.

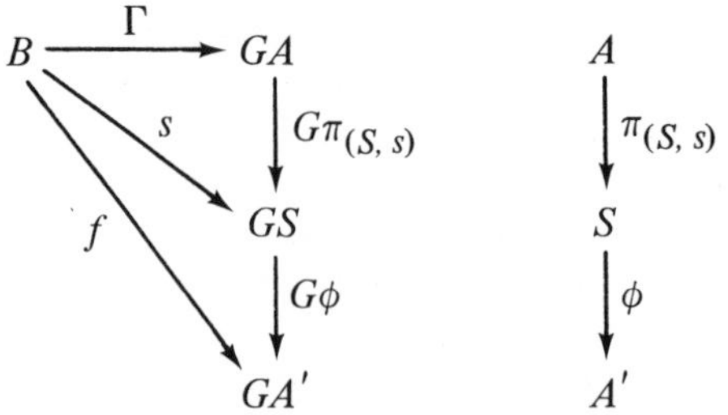

Then if $\psi : A \longrightarrow A'$ is defined to be $\phi \cdot \pi_{(S,\, s)}$, we have $G\psi \cdot \Gamma = f$ as desired.

Notice that if $\Gamma' : B \longrightarrow GA'$ and $h : A' \longrightarrow A$ satisfy

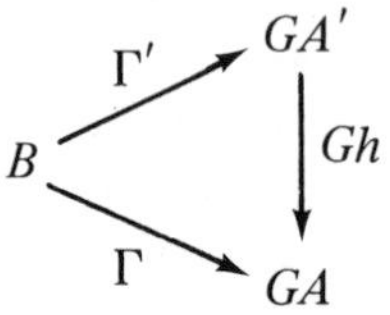

then (A', Γ') is also a one-element solution set for B. Having established that there exists a one-element solution set (A, Γ) we now construct a new one (A', Γ') as above, 'sufficiently far back' to guarantee the uniqueness property as well. [Recall our discussion (following 7.2.4) of a free object being *initial* in a suitable category.]

To this end, fix (A, Γ) and consider the set **X** of all **A**-morphisms $x : A \longrightarrow A$ with the property that $Gx \cdot \Gamma = \Gamma$. Let $(p_x : P \longrightarrow A)$ be the product, indexed by the elements of **X**, of copies of A. Define $\alpha, \beta : A \longrightarrow P$ by $p_x \cdot \alpha = \mathrm{id}_A$ for all x and $p_x \cdot \beta = x$ for all x. Let

$$i : FB \longrightarrow A = \mathrm{eq}(\alpha, \beta) \text{ in } \mathbf{A}. \tag{1}$$

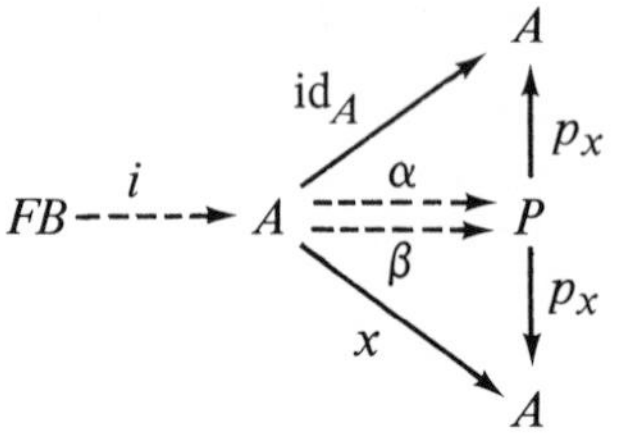

By the hypothesis on G, $(Gp_x : GP \longrightarrow GA)$ is a product in **B**.

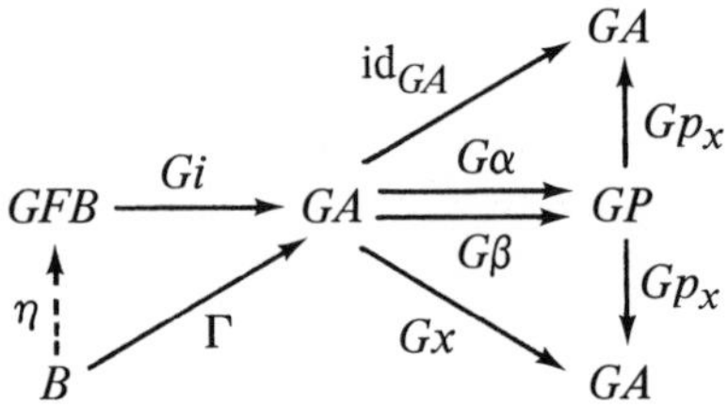

Recalling that each $x \in \mathbf{X}$ satisfies $Gx \cdot \Gamma = \Gamma$, we have

$$Gp_X \cdot G\alpha \cdot \Gamma = G(\mathrm{id}_A) \cdot \Gamma = \Gamma = Gx \cdot \Gamma = Gp_X \cdot G\beta \cdot \Gamma.$$

Therefore, $G\alpha \cdot \Gamma = G\beta \cdot \Gamma$ and so $Gi = \mathrm{eq}(G\alpha, G\beta)$ in $\mathbf{B}$ gives us a unique $\eta : B \longrightarrow GFB$ such that $Gi \cdot \eta = \Gamma$.

As we remarked earlier, (FB, η) surely satisfies the existence criterion in the universal property defining 'freeness'. We will show that (FB, η) also satisfies the uniqueness criterion and is, thereby, the free $\mathbf{A}$-object over B with respect to G:

Suppose that $\psi_1, \psi_2 : FB \longrightarrow A'$ are arbitrary $\mathbf{A}$-morphisms out of FB such that $G\psi_1 \cdot \eta = G\psi_2 \cdot \eta$. We must show that $\psi_1 = \psi_2$. Set

$$j : E \longrightarrow FB = \mathrm{eq}(\psi_1, \psi_2) \tag{2}$$

in $\mathbf{A}$. Then $Gj = \mathrm{eq}(G\psi_1, G\psi_2)$ in $\mathbf{B}$.

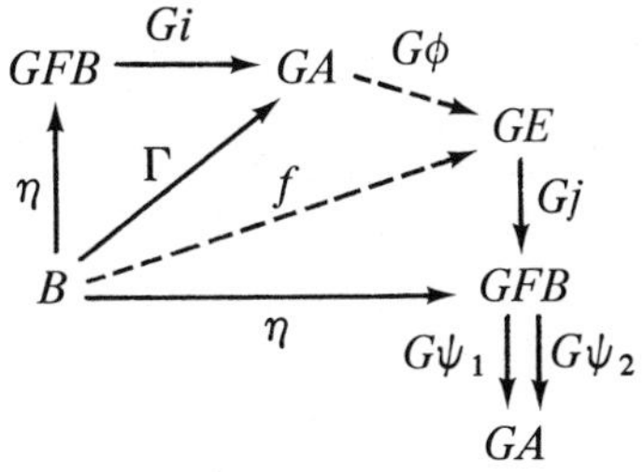

Since $G\psi_1 \cdot \eta = G\psi_2 \cdot \eta$ by hypothesis, there exists unique $f : B \longrightarrow GE$ in $\mathbf{B}$ such that $Gj \cdot f = \eta$. As (A, Γ) is a one-element solution set there exists $\phi : A \longrightarrow E$ with $G\phi \cdot \Gamma = f$. Define $x : A \longrightarrow A$ in $\mathbf{A}$ by

$$x = A \xrightarrow{\phi} E \xrightarrow{j} FB \xrightarrow{i} A .$$

Then $Gx \cdot \Gamma = Gi \cdot Gj \cdot G\phi \cdot \Gamma = Gi \cdot Gj \cdot f = Gi \cdot \eta = \Gamma$, so $x \in \mathbf{X}$. Recalling (1) that $i = \mathrm{eq}(\alpha, \beta)$, and recalling the diagram defining α and β we have that $x \cdot i = p_X \cdot \beta \cdot i = p_X \cdot \alpha \cdot i = i$. Since x was defined to be $i \cdot j \cdot \phi$ we have $i \cdot j \cdot \phi \cdot i = i$. But $i = i \cdot \mathrm{id}_{FB}$ and i is a monomorphism (2.3.6), so that $j \cdot \phi \cdot i = \mathrm{id}_{FB}$. But, by (2),

$$\psi_1 = \psi_1 \cdot \mathrm{id}_{FB} = \psi_1 \cdot j \cdot \phi \cdot i = \psi_2 \cdot j \cdot \phi \cdot i = \psi_2 .$$

Thus (FB, η) satisfies the uniqueness condition and so is indeed free over B with respect to G. □

Now that the proof of the adjoint functor theorem is complete let us re-open our discussion on the 'size' of the solution set $\mathbf{S}$. The only crucial use of $\mathbf{S}$ was to form the product $A = \Pi(S \mid (S, s) \in \mathbf{S})$. Hence one property of 'smallness' for sets is that "**A** has products" means any *small* set of $\mathbf{A}$-objects has a product. As we saw, in the category **Set**, 'large' products do not exist. (For example, the product of *all* non-empty sets does not exist.) It was important that the set $\mathbf{X}$ in the proof was sufficiently small so that the product of **X** copies of A existed in **A**. Actually most categories **A have small hom sets** (or are **locally small**) in the sense that $\mathbf{A}(A_1, A_2)$ is a small set for all A_1 and A_2 in **A**. For our purposes regarding **X**, it was surely enough to ask this whenever $A_1 = A_2$.

"Sufficiently small" can mean very small. We saw in 4.2.2 that **Met** has finite products, but not infinite ones. The proof of the adjoint functor theorem can be adapted for functors $\mathbf{Met} \longrightarrow \mathbf{K}$ if the solution sets **S** are required to be finite.

Exercises

1 Let $\mathbf{Set}_*$ be the category of pointed sets defined following 7.1.3 and let $G : \mathbf{Set}_* \longrightarrow \mathbf{Set}$ be the underlying set functor. Use the adjoint functor theorem to prove that G has a left adjoint. Then construct the free objects directly.

2 A poset $(X, \leqslant)$ is *linearly ordered* if for every x, y in X, either $x \leqslant y$ or $y \leqslant x$. Let **Loset** be the category of linearly ordered posets and order-preserving maps and let $G : \mathbf{Loset} \longrightarrow \mathbf{Poset}$ be the forgetful functor. Prove that G satisfies the solution set condition. [Hint: A natural solution set for the poset $(X, \leqslant)$ is $\{(X/R, \leqslant', s) \mid R$ is an equivalence relation on X, $\leqslant'$ is a linear ordering on X/R, and $s : (X, \leqslant) \longrightarrow (X/R, \leqslant')$ is order-preserving$\}$.] If $X = \{a, b, c\}$, $\leqslant = \{(a, a), (b, b), (c, c), (a, b), (a, c)\}$ show that no free loset over $(X, \leqslant)$ with respect to G exists. [Hint: Consider the consequence of order-preserving maps from $(X, \leqslant)$ to $(X, \leqslant')$ where $\leqslant' = \leqslant \cup \{(b, c)\}$.]

3 Use the adjoint functor theorem to prove that the forgetful functor $G : \mathbf{Vect} \longrightarrow \mathbf{Grp}$ has a left adjoint. [Hint: For a fixed group H consider the factorization $f : H \longrightarrow GV = G(\text{inc}_{V'}) \cdot s : H \longrightarrow GV' \longrightarrow GV$ where V' is the vector subspace generated by $f(H)$ in V.]

4 A group is **Boolean** if $x^2 = e$ for all x. Use the adjoint functor theorem to prove that the forgetful functor from Boolean groups and homomorphisms to **Grp** has a left adjoint. Similarly, prove that the forgetful functor from Boolean groups to **Set** has a right adjoint.

5 Prove that a functor $\mathbf{Met} \longrightarrow \mathbf{K}$ which preserves finite products and collective equalizers and for which every **K**-object has a finite solution set, has a left adjoint. Here, given a family $(f_i : A \longrightarrow B \mid i \in I)$, a **collective equalizer of** (f_i) is a morphism $g : E \longrightarrow A$ such that $f_i \cdot g = f_j \cdot g$ for all i, j in I and such that whenever $f_i \cdot h = f_j \cdot h$ there exists unique ψ with $\psi \cdot g = h$. [Hint: Prove *directly* that the class **X** of the proof of 2 has a collective equalizer, thus avoiding the use of **X**-indexed products.]

Chapter 9

MONOIDAL AND CLOSED CATEGORIES

As we saw in the passage from Chapter 1 to Chapter 2, much of elementary category theory has consisted in the search for the right way to speak about sets and maps so that the statement of their properties will carry over to many other domains of discourse. In Section 1 we present some considerations from set theory which have led category theorists to study closed categories, monoidal categories, and toposes. We shall then discuss monoidal categories at greater length in the later sections.

9.1 MOTIVATION FROM SET THEORY

The first property of **Set** that we now wish to capture is that it 'supports' the construction of monoids. To see what this means, let us rephrase the definition, 3.1.2, of a monoid in arrow-form denoting the multiplication as $m : M \times M \longrightarrow M$. Noting that there is an isomorphism

$$\alpha : M \times (M \times M) \longrightarrow (M \times M) \times M : (a, (b, c)) \mapsto ((a, b), c)$$

in the category **Set** (in fact, such an isomorphism exists in any category with binary products: see exercise 2.4.1), we may re-express the associative law in the form

$$\begin{array}{ccc} & (M \times M) \times M \xrightarrow{m \times \mathrm{id}} & M \times M \\ \overset{\alpha}{\nearrow} & & \\ M \times (M \times M) & & \Big\downarrow m \\ \Big\downarrow \mathrm{id} \times m & & \\ M \times M & \xrightarrow[m]{\qquad\qquad} & M \end{array}$$

Again, letting 1 denote any convenient one-element set (which is thus terminal), the identity element may be represented as a map $e : 1 \longrightarrow M$. That e is a two-sided identity may then be expressed in the form

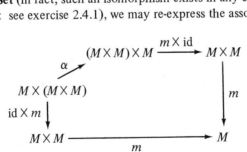

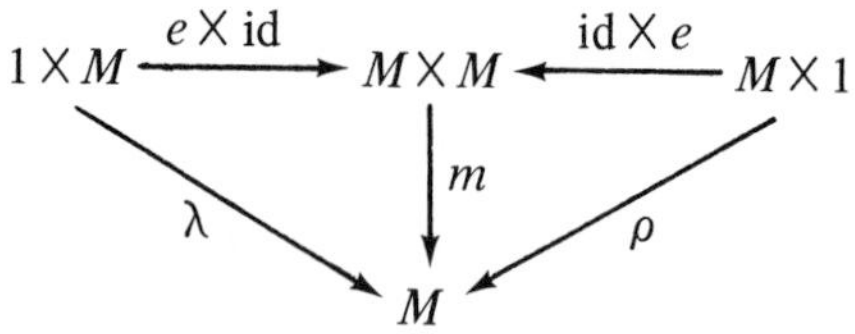

where $\lambda : 1 \times M \longrightarrow M : (1, m) \mapsto m$ and $\rho : M \times 1 \longrightarrow M : (m, 1) \mapsto m$ are isomorphisms.

Generalizing this construction immediately yields:

1 **DEFINITION**: Let **K** be an arbitrary category which possesses binary products $A \times B$ and a terminal object 1. **A monoid in the category K** is an object M of **K** equipped with morphisms $m : M \times M \longrightarrow M$ (multiplication) and $e : 1 \longrightarrow M$ (identity) subject to the requirement that the following diagrams are commutative,

associative law

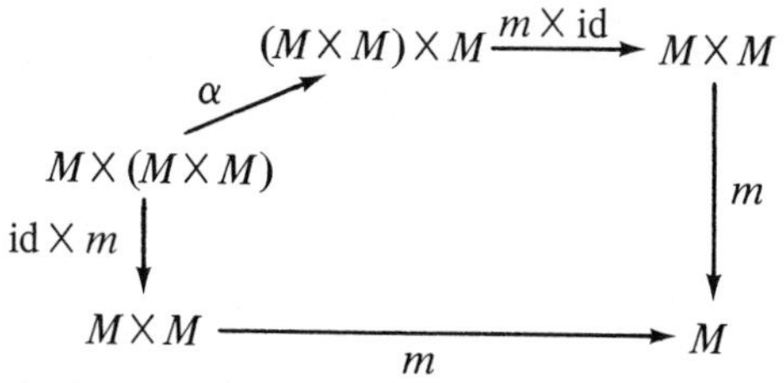

e **is a two-sided unit**

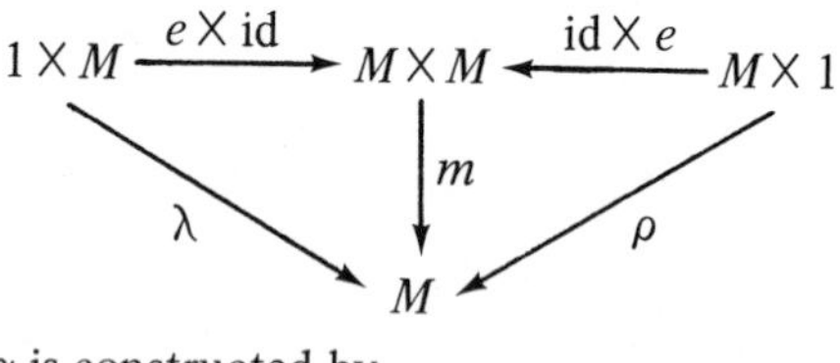

where the isomorphism α is constructed by

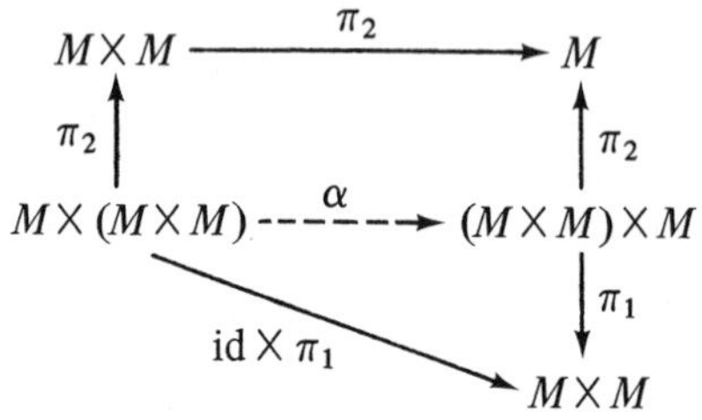

and $\lambda = \pi_2$, $\rho = \pi_1$ are projections. α, λ and ρ are isomorphisms (see exercise 1).

Thus every monoid in the sense of 3.1.2 is a monoid in **Set**. In the category **Vect**, every vector space is itself a monoid: just note that the above diagrams are satisfied by multiplication $+ : X \times X \longrightarrow X$ and zero for unit $0 \longrightarrow X$ (in **Vect**, 1 is also initial and most naturally written as '0').

2 **PROPOSITION**: The monoids in **Mon** are the abelian monoids.

Proof: Monoids in **Mon** are monoids M equipped with homomorphisms $m' : M \times M \longrightarrow M$ and an element $e' : 1 \longrightarrow M$ which satisfy the diagram of 1. First note that $m' : X \times X \longrightarrow X$ is a monoid homomorphism if and only if m' satisfies

$$m'(e, e) = e$$

$$m'(xa, yb) = m'(x, y)\, m'(a, b).$$

If $e' : 1 \longrightarrow M$ is a monoid homomorphism then we have at once that $e = e'$ (where we identify elements and functions from a one-element set in the obvious way). If (M, m', e) is a monoid in **Mon** then also

$$m'(t, m'(u, v)) = m'(m'(t, u), v)$$

$$m'(e, x) = x = m'(x, e)$$

But then $xy = m'(x, e)\, m'(e, y) = m'(xe, ey) = m'(x, y)$ and m coincides with the original multiplication. Moreover, $xy = m(ex, ye) = m(e, y)\, m(x, e) = yx$. As we saw in Section 5.2, monoids satisfying $xy = yx$ are called *abelian* or *commutative*. Conversely, if X is an abelian monoid, X is easily seen to be a monoid in **Mon**. □

The idea of a *monoidal category*, then, is that we can draw diagrams like those in 1 above, but that we may replace the functor $\times : \mathbf{K} \times \mathbf{K} \longrightarrow \mathbf{K}$ by some other functor $\otimes : \mathbf{K} \times \mathbf{K} \longrightarrow \mathbf{K}$ which has natural transformations α, λ and ρ available with suitable properties. We defer the details to Definition 2.1 – here we give an example of an interesting $\otimes$ [the reader versed in multilinear algebra will recognize it as a variant on the usual tensor product; where *separately homomorphic* corresponds to the usual expression *bilinear*]:

3 **EXAMPLE**: Let A, B, C be abelian monoids. A function $f : A \times B \longrightarrow C$ is **separately homomorphic** if for each $a \in A$, $b \in B$ the maps

$$B \longrightarrow C, \; x \mapsto f(a, x)$$

and

$$A \longrightarrow C, \; x \mapsto f(x, b)$$

are monoid homomorphisms. A **tensor product of** A, B is a pair $(A \otimes B, \eta)$ where $A \otimes B$ is a monoid and $\eta : A \times B \longrightarrow A \otimes B$ is separately homomorphic with the universal property that

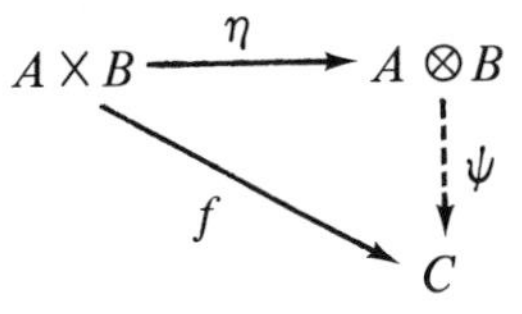

for every separately homomorphic $f : A \times B \longrightarrow C$ there exists a unique monoid homomorphism $\psi : A \otimes B \longrightarrow C$ such that $\psi\eta = f$.

It is clear that the tensor product is unique up to isomorphism if it exists. It is possible to use the adjoint functor theorem to prove $A \otimes B$ exists (see exercise 2.5).

In our discussion of right adjoints, we saw following 7.2.8 that for any set A, the functor $- \times A$: **Set** $\longrightarrow$ **Set** has a right adjoint $(-)^A$ which sends B to the set B^A of all maps from A to B. In fact, we saw that to every $f : A' \times A \longrightarrow B$ there corresponds a unique $f^{\boldsymbol{\cdot}} : A' \longrightarrow B^A$

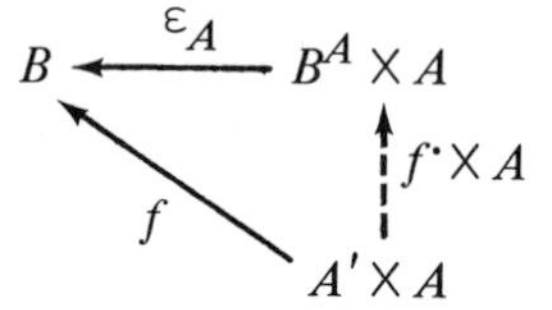

such that $\varepsilon_A(f^{\boldsymbol{\cdot}}(a'), a) = f^{\boldsymbol{\cdot}}(a')(a) = f(a', a)$: namely $f^{\boldsymbol{\cdot}}(a') = f(a', \cdot)$. We shall briefly discuss in Section 9.3 the suggestive fact that B^A may also be written as **Set**(A, B), but here we content ourselves with noting that **Set** is the motivating example of a *Cartesian closed category*:

4 **DEFINITION**: A category **K** is **cartesian closed** if it has finite products, and if the functor $- \times A : \mathbf{K} \longrightarrow \mathbf{K}$ has a right adjoint for every object A of **K**.

We close this section by noting that much category theory is now based on the familiar property of set theory that there is a bijection linking subsets of a set X with their characteristic functions:

Let $\Omega = \{t, f\}$, say, be a fixed two-element set. Then given any subset A of a set X, we define its **characteristic function**

$$\chi_A : X \longrightarrow \Omega \qquad \text{where} \qquad \chi_A(x) = \begin{cases} t & \text{if } x \in A \\ f & \text{if } x \notin A. \end{cases}$$

Conversely, given any map $f : X \longrightarrow \Omega$, it defines the subset $f^{-1}(t)$ of X, and it is clear that the passages $A \leftrightarrow \chi_A$ and $f \leftrightarrow f^{-1}(t)$ are mutually inverse. Now if we identify t with the map $t : 1 \longrightarrow \Omega$, we note that the following is a pullback:

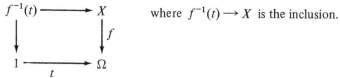

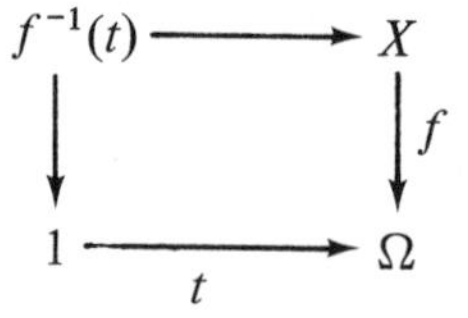

where $f^{-1}(t) \longrightarrow X$ is the inclusion.

For suppose that we have

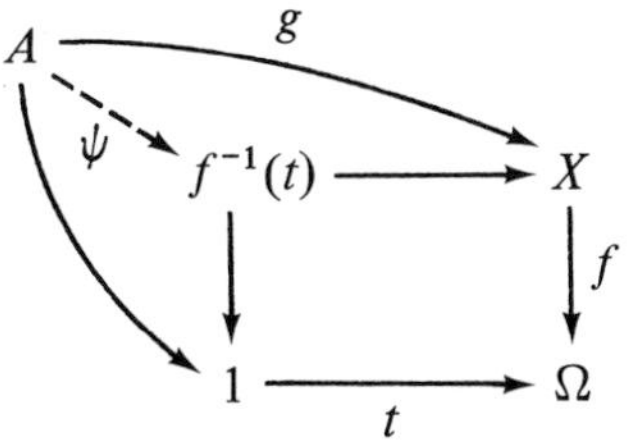

Then the unique ψ is defined by $\psi(a) = g(a)$, since $f(g(a)) = t$ implies that $g(a)$ is in $f^{-1}(t)$.

Thus **Set**, equipped with the map $t : 1 \longrightarrow \Omega$, is an example of a topos:

5   DEFINITION: A **topos** is a cartesian closed category **V** equipped with a **subobject classifier** (Ω, t); i.e., $t : 1 \longrightarrow \Omega$ is such that for any $f : X \longrightarrow \Omega$, the pullback

$$
\begin{array}{ccc}
f^{-1}(t) & \longrightarrow & X \\
\downarrow & & \downarrow f \\
1 & \xrightarrow{\ t\ } & \Omega
\end{array}
$$

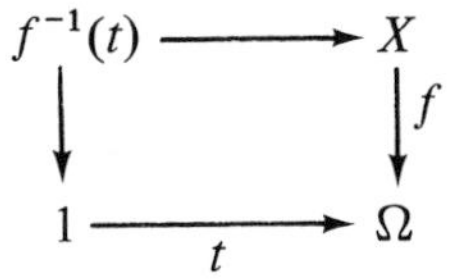

exists, with $f^{-1}(t) \longrightarrow X$ (which is a monomorphism since $t : 1 \longrightarrow \Omega$ is a monomorphism [why?] and pullbacks preserve monomorphisms), in such a way that the passage from f to $f^{-1}(t)$ is a bijection between $\mathbf{V}(X, \Omega)$ and mono-subobjects of X.

The importance of this concept is that it can be applied to other "categories like **Set**" where Ω is "an arbitrary Heyting algebra". Logicians have used a topos as a model of intuitionistic set theory, as a setting for a proof of the independence of the continuum hypothesis; while algebraic geometers view a topos as a category of sheaves, as a generalized setting for Grothendieck topologies. But the vocabulary of these applications, let alone their details, takes us far beyond the elementary level of this book, and so we refer the reader to "Toposes, Algebraic Geometry and Logic" (F. W. Lawvere, Ed.: Springer Lecture Notes in Mathematics **274** (1972)) and the bibliography there for further information on this new research frontier.

Exercises

1   In the notation of 1, show that λ and ρ are isomorphisms with $\lambda^{-1} = (\phi, \mathrm{id})$ and $\rho^{-1} = (\mathrm{id}, \phi)$, where $\phi : M \longrightarrow 1$. Show that α is an isomorphism. [Hint: Define α^{-1} in a manner similar to α.]

2   Show that **R** – with its usual metric and additive structure – is *not* a monoid in **Met**. Show, however, that **R** is a monoid in **Top** (whose products were defined in exercise 4.3.3).

3 Show that $\mathbf{Set}^{\mathrm{op}}$ has exactly one monoid.

4 Show that "$\mathbf{N} \otimes A = A$ in **Abm**", i.e. that (A, η), where $\eta : \mathbf{N} \times A \longrightarrow A$, $(n, a) \mapsto a + \ldots + a$ (n times), is a tensor product of $\mathbf{N}$, A. [Hint: Given separately homomorphic $f : \mathbf{N} \times A \longrightarrow B$, define $\psi(a) = f(1, a)$.] Show that if the monoid A is not abelian then η is not separately homomorphic.

5 Prove that any two subobject classifiers (Ω, t), (Ω', t') of a topos are isomorphic. [Hint: Apply the "pullback-pasting" exercise 2.4.9.]

6 Prove that $\mathbf{Dyn}(X_0)$ of Section 6.3 is a topos. [Hints: $(Q_1, \delta_1) \times (Q_2, \delta_2) = (Q_1 \times Q_2, \delta)$ where $\delta((q_1, q_2), x) = (\delta_1(q_1, x), \delta_2(q_2, x))$. Ω is the set of all subsets $S \subset X_0{}^*$ with that property that $sw \in S$ for all $s \in S$, $w \in X_0{}^*$, with dynamics

$$\Omega \times X_0 \longrightarrow \Omega, \quad (S, x) \mapsto \{w \mid wx \in S\};$$

$t = X_0^*$ (check that $t : 1 \longrightarrow \Omega$ is a dynamorphism!), and the subdynamics (A, δ_0) of (Q, δ) corresponds to the dynamorphism

$$\chi_A : (Q, \delta) \longrightarrow \Omega, \quad q \mapsto \{w \in X_0{}^* \mid q\delta^*(q, w) \in A\}.$$

Define $(R, \gamma)^{(Q, \delta)}$ to be the subdynamics of all dynamorphisms.]

9.2 MONOIDS IN A MONOIDAL CATEGORY

The previous section motivates the following definition:

1 **DEFINITION**: A **monoidal category** is a 6-tuple $(\mathbf{V}, \otimes, I, \alpha, \lambda, \rho)$ such that $\mathbf{V}$ is a category, $\otimes : \mathbf{V} \times \mathbf{V} \longrightarrow \mathbf{V}$ is a functor, I is an object of $\mathbf{V}$ and† $\alpha_{A,B,C} : A \otimes (B \otimes C) \longrightarrow (A \otimes B) \otimes C$, $\lambda_A : I \otimes A \longrightarrow A$, $\rho_A : A \otimes I \longrightarrow A$ are natural equivalences of functors $\mathbf{V} \times \mathbf{V} \times \mathbf{V} \longrightarrow \mathbf{V}$, $\mathbf{V} \longrightarrow \mathbf{V}$, $\mathbf{V} \longrightarrow \mathbf{V}$, such that the following three **coherence axioms** hold:

$$\begin{array}{ccccc}
A \otimes (B \otimes (C \otimes D)) & \xrightarrow{\alpha_{A,B,C \otimes D}} & (A \otimes B) \otimes (C \otimes D) & \xrightarrow{\alpha_{A \otimes B,C,D}} & ((A \otimes B) \otimes C) \otimes D \\
\Big\downarrow \mathrm{id}_A \otimes \alpha_{B,C,D} & & & & \Big\uparrow \alpha_{A,B,C} \otimes \mathrm{id}_D \\
A \otimes ((B \otimes C)) \otimes D) & & \xrightarrow[\alpha_{A,B \otimes C,D}]{\hspace{6em}} & & (A \otimes (B \otimes C)) \otimes D
\end{array}$$

$$I \otimes I \xrightarrow{\lambda_I} I \; = \; I \otimes I \xrightarrow{\rho_I} I$$

---

† For readability, we now write the object maps for our natural transformations as $\alpha_{A,B,C}$, λ_A and ρ_A, rather than $\alpha(A, B, C)$, λA, and ρA respectively.

$$\begin{array}{ccc} A\otimes(I\otimes B) & \xrightarrow{\alpha_{A,I,B}} & (A\otimes I)\otimes B \\ & {\scriptstyle \mathrm{id}_A\otimes\lambda_B}\searrow \quad \swarrow{\scriptstyle \rho_A\otimes \mathrm{id}_B} & \\ & A\otimes B & \end{array}$$

We shall often refer to the monoidal category simply as **V**, rather than specifying $\otimes$, I, α, λ and ρ explicitly.

We say that **V** is **strictly monoidal** if, further, α, λ and ρ are all equalities.

In **Set** we may check that, e.g., $((A \times 1) \times ((B \times C) \times 1) \times D$ is isomorphic to $(A \times B) \times (C \times D)$. The purpose of the coherence axioms in a monoidal category is to guarantee that two such expressions involving $\otimes$, I which 'should be isomorphic' are in fact isomorphic via a unique expression built up from iterated use of α, λ and ρ. [For a precise statement and the rather difficult proof consult Mac Lane's "Categories for the Working Mathematician", Section VII.2.] We then generalize 1.1 to Definition 2, and verify that this is indeed a generalization in Theorem 3.

2 **DEFINITION**: **A monoid in the monoidal category** $(\mathbf{V}, \otimes, I, \alpha, \lambda, \rho)$ is a triple (X, m, e) where $m : X\otimes X \longrightarrow X$ and $e : I \longrightarrow X$ are subject to the **associative law**

$$\begin{array}{ccc} & (X\otimes X)\otimes X & \xrightarrow{m\otimes \mathrm{id}} & X\otimes X \\ & {\scriptstyle \alpha_{X,X,X}}\nearrow & & \\ X\otimes(X\otimes X) & & & \Big\downarrow {\scriptstyle m} \\ {\scriptstyle \mathrm{id}\otimes m}\Big\downarrow & & & \\ X\otimes X & & \xrightarrow[m]{\qquad\qquad} & X \end{array}$$

and the **unitary laws**

$$\begin{array}{ccccc} I\otimes X & \xrightarrow{e\otimes \mathrm{id}} & X\otimes X & \xleftarrow{\mathrm{id}\otimes e} & X\otimes I \\ & {\scriptstyle \lambda_X}\searrow & \Big\downarrow{\scriptstyle m} & \swarrow{\scriptstyle \rho_X} & \\ & & X & & \end{array}$$

A **monoid homomorphism** $f : (X, m, e) \longrightarrow (X', m', e')$ is a **V**-morphism $f : X \longrightarrow X'$ which satisfies

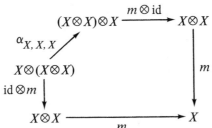

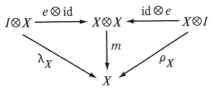

$$\begin{array}{ccc} X \otimes X & \xrightarrow{f \otimes f} & X' \otimes X' \\ {\scriptstyle m}\downarrow & & \downarrow{\scriptstyle m'} \\ X & \xrightarrow[f]{} & X' \end{array}$$

3 **PROPOSITION**: Let **K** be a category with finite products including a terminal object 1. Then $(\mathbf{K}, \times, 1, \alpha, \lambda, \rho)$ is a monoidal category for the 'obvious' α, λ, ρ.

Proof Construction: $\times : \mathbf{K} \times \mathbf{K} \longrightarrow \mathbf{K}$ becomes a functor by choosing arbitrary but definite products $A \longleftarrow A \times B \longrightarrow B$ for each pair of objects (A, B) and then defining the action on morphisms by

$$\begin{array}{ccc} A & \xrightarrow{f} & B \\ {\scriptstyle \pi_1}\uparrow & & \uparrow{\scriptstyle \pi_1} \\ A \times C & \xrightarrow{f \times g} & B \times D \\ {\scriptstyle \pi_2}\downarrow & & \downarrow{\scriptstyle \pi_2} \\ C & \xrightarrow[g]{} & D \end{array}$$

α, λ and ρ are defined by

$$\begin{array}{ccc} B \times C & \xrightarrow{\pi_2} & C \\ {\scriptstyle \pi_2}\uparrow & & \uparrow{\scriptstyle \pi_2} \\ A \times (B \times C) & \xrightarrow{\alpha_{A,B,C}} & (A \times B) \times C \\ & \searrow^{\mathrm{id} \times \pi_1} & \downarrow{\scriptstyle \pi_1} \\ & & A \times B \end{array}$$

$$1 \times A \xrightarrow{\lambda_A} A = \pi_2$$

$$A \times 1 \xrightarrow{\rho_A} A = \pi_1$$

The remaining details are tedious but routine (cf. exercise 1.1). □

A typical application of 3 is:

4 **EXAMPLE**: A monoid in $(\mathbf{Top}, \times, 1, \alpha, \lambda, \rho)$ is a topological space (X, τ) which is also an ordinary monoid (X, m, e) such that $m : (X, \tau) \times (X, \tau) \longrightarrow (X, \tau)$ is continuous (note: $e : 1 \longrightarrow (X, \tau)$ is always continuous). This is called a **(jointly continuous) topological monoid**. A description of $(X, \tau) \times (X, \tau)$, the

well-known product space, appeared in exercise 4.3.3.

5 **EXAMPLE**: Let **V** be the poset $(\mathbf{R}^+, \geqslant)$ of nonnegative reals qua category. Then $\mathbf{V} \times \mathbf{V}$ is also a poset and

$$\otimes : \mathbf{V} \times \mathbf{V} \longrightarrow \mathbf{V} \qquad \text{defined by} \qquad x \otimes y = x + y$$

is order-preserving (i.e. $x \geqslant a$ and $y \geqslant b$ implies $x + y \geqslant a + b$), hence is a functor. Then $(\mathbf{V}, +, 0, \alpha, \lambda, \rho)$, where α, λ, ρ are identity maps, is a strictly monoidal category. A monoid in **V** amounts to a number x satisfying $x + x \geqslant x$ (existence of m) and $0 \geqslant x$ (existence of e). Thus 0 is the only monoid.

6 **EXAMPLE**: Recall the tensor product $\otimes$ of Example 1.3: Then $(\mathbf{Abm}, \otimes, I, \alpha, \lambda, \rho)$ is a monoidal category where I is the monoid $(\mathbf{N}, +, 0)$ of natural numbers and α, λ, ρ are induced by the universal property of $A \otimes B$ as indicated in the diagrams below:

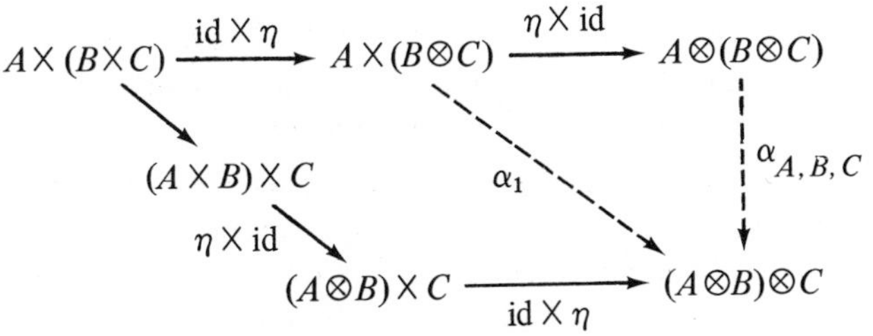

(α_1 is obtained by fixing variables in A to yield separately homomorphic maps $B \times C \longrightarrow (A \otimes B) \otimes C$.)

The remaining details that this is monoidal are numerous but routine (see exercise 6).

A monoid in **Abm** is an abelian monoid $(X, +, 0)$ which is also an ordinary monoid $(X, \cdot, 1)$ (Note: I is the free monoid on one generator so that maps $I \longrightarrow X$ are in bijective correspondence with elements of X in essentially the same way as in 5.1.2) such that $\cdot : X \times X \longrightarrow X$ is separately homomorphic, i.e. satisfies the distributive laws $x \cdot (y + z) = (x \cdot y) + (x \cdot z)$, $(x + y) \cdot z = (x \cdot z) + (y \cdot z)$. Such an $(X, +, 0, \cdot, 1)$ is known as a **semiring** in the literature.

7 **EXAMPLE**: Let **Cat** be the category whose objects are categories and whose morphisms are functors. Then

$$\mathbf{Cat} \times \mathbf{Cat} \xrightarrow{X} \mathbf{Cat}$$
$$\mathbf{K}, \mathbf{L} \longmapsto \mathbf{K} \times \mathbf{L}$$

is a functor; given $F : \mathbf{K} \longrightarrow \mathbf{A}$, $G : \mathbf{L} \longrightarrow \mathbf{B}$ we have $F \times G : \mathbf{K} \times \mathbf{L} \longrightarrow \mathbf{A} \times \mathbf{B}$ defined by $(F \times G)(K, L) = (FK, GL)$ and, for $f : K \longrightarrow K'$, $g : L \longrightarrow L'$, $(F \times G)(f, g) = (Ff, Gg)$. Let **1** be the one-object one-morphism category. Then defining α, λ, ρ in the 'obvious' way (the details are reminiscent of **Set**), $(\mathbf{Cat}, \times, 1, \alpha, \lambda, \rho)$ is a monoidal category. A monoid in **Cat** is a strictly monoidal category.

Exercises

1 Verify that the coherence axioms hold for $(\mathbf{Set}, \times, 1, \alpha, \lambda, \rho)$.

2 Show that monoids in a monoidal category form a category.

3 Write out the details of the proof of 3.

4 Show that every abelian monoid M – qua one-object category – is strictly monoidal with $\otimes$ = monoid addition = category composition, and that a monoid in M is a pair $(a, b) \in M \times M$ with $ab = e$.

5 Let A, B be fixed abelian monoids. Define a functor $U : \mathbf{Abm} \longrightarrow \mathbf{Set}$ as follows. $U(C) = \{f \mid f \text{ is separately homomorphic } A \times B \longrightarrow C\}$. For $g : C \longrightarrow D \in \mathbf{Abm}$, let $U(g) : U(C) \longrightarrow U(D)$, $f \mapsto gf$. Use the adjoint functor theorem to prove that U has a left adjoint. Conclude that the tensor product of A, B exists, being the free object over 1 with respect to U.

6 Supply the remaining details of 6. (See exercise 1.4.)

7 Re-do 6 replacing **Abm** with the category of abelian groups. Here $I = \mathbf{Z}$, the group of all integers.

8 Let **V** be the category of abelian groups and let $\otimes, I, \alpha, \lambda, \rho$ be as in exercise 7. Show that $(\mathbf{V}, \otimes, I, -\alpha, \lambda, \rho)$ is not a monoidal category because the coherence axioms fail.

9 A **rewrite system** is a pair (Σ, S) where Σ is a set and $S \subset \Sigma^* \times \Sigma^*$ where Σ^* is the free monoid generated by Σ. Write $w \Rightarrow w'$ if $w = w_1 u w_2$, $w' = w_1 v w_2$ and $(u, v) \in S$ for some w_1, u, v, w_2. Let $\leqslant$ be the intersection of all reflexive and transitive relations on Σ^* containing $\Rightarrow$. Set $\mathbf{V} = (\Sigma^*, \leqslant)$ qua category, define $w \otimes w' = ww'$, let I be the empty word and let α, λ, ρ be identity transformations. Prove that $(\mathbf{V}, \otimes, I, \alpha, \lambda, \rho)$ is a monoidal category. [D. B. Benson ("An abstract machine theory for formal language parsers," *Acta Informatica*, **3** (1974), 187-202) has used this construction for parsing formal languages.]

9.3 CATEGORIES OVER A MONOIDAL CATEGORY

Monoidal categories provide not only the setting for defining monoids (i.e. one-object categories) but arbitrary categories:

1 **DEFINITION**: Let $\mathbf{V} = (\mathbf{V}, \otimes, I, \alpha, \lambda, \rho)$ be a monoidal category. A **V-category A** (which is not *a priori* a category) is defined by the following data and axioms (obtained from 2.2.1 by replacing **Set** with **V**):

We are given a class obj(**A**) of **objects** of **A**. For each pair (A, B) of **A**-objects we are given an object $\mathbf{A}[A, B]$ of **V**. For each triple (A, B, C) of **A**-objects we are given a **V**-morphism

$$\mathbf{A}[A, B] \otimes \mathbf{A}[B, C] \xrightarrow{c_{A,B,C}} \mathbf{A}[A, C]$$

of **internal composition.** For each **A**-object A we are given a **V**-morphism

$$e_A : I \longrightarrow \mathbf{A}[A, A]$$

called **internal identity**. The three axioms imposed assert that composition is associative and that e_A acts as left and right identity:

$$\begin{array}{ccc}
(\mathbf{A}[A,B]\otimes\mathbf{A}[B,C])\otimes\mathbf{A}[C,D] & \xrightarrow{c_{A,B,C}\otimes \mathrm{id}} & \mathbf{A}[A,C]\otimes\mathbf{A}[C,D] \\
\uparrow{\scriptstyle \alpha} & & \\
\mathbf{A}[A,B]\otimes(\mathbf{A}[B,C]\otimes\mathbf{A}[C,D]) & & \downarrow{\scriptstyle c_{A,C,D}} \\
{\scriptstyle \mathrm{id}\otimes c_{B,C,D}}\downarrow & & \\
\mathbf{A}[A,B]\otimes\mathbf{A}[B,D] & \xrightarrow[c_{A,B,D}]{} & \mathbf{A}[A,D]
\end{array}$$

$$\begin{array}{ccc}
I\otimes\mathbf{A}[A,B] & \xrightarrow{e_A\otimes \mathrm{id}} & \mathbf{A}[A,A]\otimes\mathbf{A}[A,B] \\
& {\scriptstyle \lambda}\searrow & \downarrow{\scriptstyle c_{A,A,B}} \\
& & \mathbf{A}[A,B]
\end{array}$$

$$\begin{array}{ccc}
\mathbf{A}[A,B]\otimes I & \xrightarrow{\mathrm{id}\otimes e_B} & \mathbf{A}[A,B]\otimes\mathbf{A}[B,B] \\
& {\scriptstyle \rho}\searrow & \downarrow{\scriptstyle c_{A,B,B}} \\
& & \mathbf{A}[A,B]
\end{array}$$

Thinking of **Set** as a monoidal category via 2.3, a **Set**-category is just an ordinary category as in 2.2.1. Also, a monoid in **V** is just a one-object **V**-category (compare 2.2 and 1).

A **V**-category "is" a category as follows: Define $|\mathbf{A}|$, the **underlying category of A**, by

$$\mathrm{obj}(|\mathbf{A}|) = \mathrm{obj}(\mathbf{A})$$

$$|\mathbf{A}|(A, B) = \mathbf{V}(I, \mathbf{A}[A, B])$$

$$\begin{array}{ccccc} |\mathbf{A}|(A, B) & \times & |\mathbf{A}|(B, C) & \longrightarrow & |\mathbf{A}|(A, C) \\ f & , & g & \mapsto & gf \end{array}$$

where

2

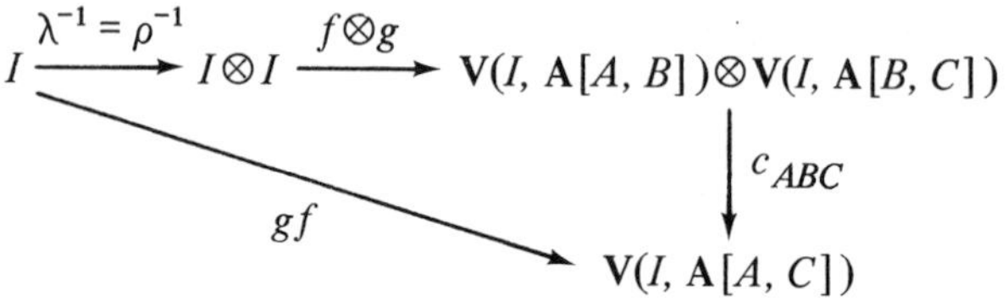

For identities use $e_A \in |\mathbf{A}|(A, A)$.

3 **PROPOSITION**: Let **V** be a monoidal category and let **A** be a **V**-category. Then $|\mathbf{A}|$, as defined above, is indeed a category.

Proof: Since the diagrams needed are of unwieldy size, let us abbreviate $\mathbf{A}[A, B]$ as AB. To prove that composition in $|\mathbf{A}|$ is associative let $f \in |\mathbf{A}|(A, B)$, $g \in |\mathbf{A}|(B, C)$, $h \in |\mathbf{A}|(C, D)$ and consult diagram 4.

4

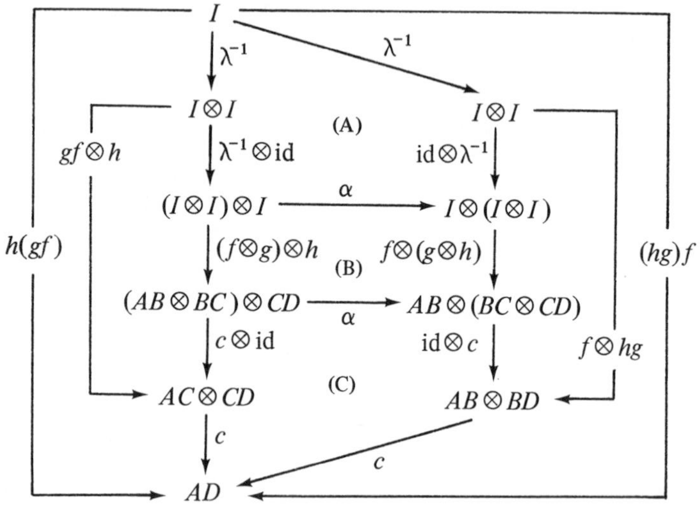

The commutativity of (A) is referred to exercise 3. (B) is the naturality of α. (C) is just the associativity law for c in the **V**-category **A**. To show that e_A is a left identity let $f \in |\mathbf{A}|(A, B)$ and consider

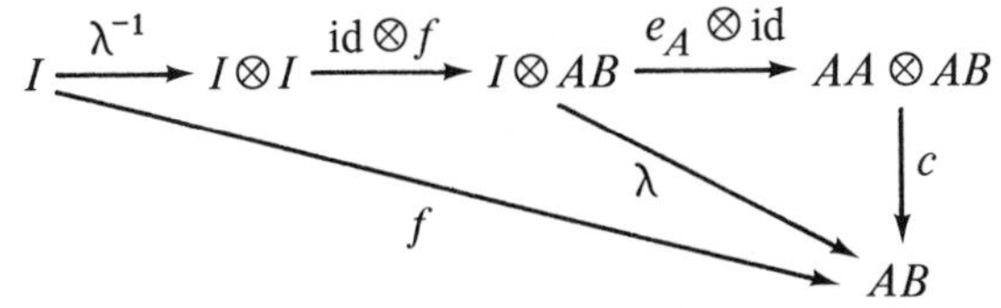

The first triangle is the naturality of λ and the second is an axiom on **A**. That e_A is a right identity is similar. □

5 **EXAMPLE**: Let **V** be the monoidal category **Abm** of 2.6. Note that a morphism $I \longrightarrow A$ in **V** corresponds to an element of (the underlying set of the abelian monoid) A (cf. 5.1.1). Thus, if **A** is a **V**-category, $|\mathbf{A}|(A, B)$ may be identified with the elements of the abelian monoid $\mathbf{A}[A, B]$. Thus the monoid homomorphism $c_{ABC} : \mathbf{A}[A, B] \otimes \mathbf{A}[B, C] \longrightarrow \mathbf{A}[A, C]$ is, as a function, composition in $|\mathbf{A}|$; since such composition is separately homomorphic we have the distributive laws

$$h(g_1 + g_2) = hg_1 + hg_2, \quad (g_1 + g_2)f = g_1 f + g_2 f.$$

Continuing in this vein it is possible to check that **V**-categories are coextensive with the **Abm**-categories of Chapter 5.

6 **EXAMPLE**: (F. W. Lawvere) Let **V** be the monoidal category of positive real numbers of 2.5. Then a **V**-category **A** is given by a set $X = \text{obj}(\mathbf{A})$ together with $d(x, y) = \mathbf{A}[x, y] \geqslant 0$ subject to the axioms

$$d(x, y) + d(y, z) \geqslant d(x, z) \qquad \text{(existence of } c_{x,y,z})$$

$$d(x, x) = 0 \qquad \text{(existence of } e_x)$$

noting that the other axioms are automatic since all diagrams in **V** commute. Thus a metric space is a **V**-category (X, d) such that, also, $d(x, y) = d(y, x)$ and $d(x, y) = 0$ implies $x = y$.

In concluding this section we mention – very briefly – some important concepts, awareness of which is within the categorical imperative.

The definition of a **V**-category suggests that **V** is a 'universe' and that one can "do all category theory inside **V**". An obvious failure of this doctrine is that **V** need not itself be a **V**-category (whereas, in ordinary category theory, **Set** is a **Set**-category). A remedy to this difficulty is to require that, for each A in **V**, $- \otimes A : \mathbf{V} \longrightarrow \mathbf{V}$ has a right adjoint (via specified adjunctions). Also, for technical reasons, **V** is required to be **symmetric**, i.e. there is a specified natural equivalence $\gamma_{A,B} : A \otimes B \longrightarrow B \otimes A$ and additional coherence axioms relating γ

to α, λ, ρ. Such **V** is called a **closed** category.

Recalling the fact that **Set** is a *cartesian closed category* (as discussed preceding 1.4) we see that **Set** is a closed category and 2.6, 2.7 and exercise 2.7 provide additional examples. (In Example 2.6, $\mathbf{V}[A, B]$ is the set of monoid homomorphisms with pointwise operations.) If **V** is closed, $\mathbf{V}[A, B]$ provides the **V**-object of morphisms for a **V**-category whose underlying category is **V**, i.e. "**V** is a **V**-category".

When **V** is closed, all important concepts of category theory can be internalized. For example, a **V-functor** $H : \mathbf{A} \longrightarrow \mathbf{B}$ where **A**, **B** are **V**-categories is specified by providing a passage $A \mapsto HA$ on objects as usual but the action on morphisms is a **V**-morphism

$$H_{AB} : \mathbf{A}[A, B] \longrightarrow \mathbf{B}[HA, HB]$$

and "preservation of composition and identities" is expressed by commutative diagrams. There are **V**-natural transformations, **V**-adjoint **V**-functors, **V**-limits and **V**-colimits. It is also possible to 'tensor' or 'cotensor' an object A in a **V**-category **A** with an object n in **V** (generalizing, respectively, the n-fold copower and n-fold power of copies of A when $\mathbf{V} = \mathbf{Set}$). **V**-notions are everywhere.

Not all important monoidal categories are closed. In Theorem 4.5 we construct an important class of non-symmetric strictly monoidal categories.

Exercises

1 Prove that a **Set**-category is just an ordinary category.

2 Prove that **V**-monoids are one-object **V**-categories.

3 Prove the commutativity of (A) in 4 using the coherence axioms for **V**. Since the failure of associativity 'on the nose' for $|\mathbf{A}|$ is unpleasant to contemplate, this underscores the importance of coherence in a monoidal category.

4 In the context of 3, prove that e_A is a right identity.

5 Fill in the details to 4.

6 Generalize the constructions of **Abm** to provide **Vect** with the structure of a closed category.

9.4 THE GODEMENT CALCULUS

The material of this section may be construed as providing most of the proof that the category **Cat** of categories and functors is a **Cat**-category. Our first definition provides $\mathbf{Cat}[\mathbf{K}, \mathbf{L}]$.

1 **DEFINITION:** Let **K**, **L** be categories. The **functor category** $\mathbf{L}^{\mathbf{K}}$ is the category whose objects are functors $F:\mathbf{K}\longrightarrow\mathbf{L}$ and whose morphisms are natural transformations $\tau:F_1 \stackrel{\cdot}{\longrightarrow} F_2$. Given $\tau:F_1 \stackrel{\cdot}{\longrightarrow} F_2$, $\sigma:F_2 \stackrel{\cdot}{\longrightarrow} F_3$, $\sigma\!\cdot\!\tau:F_1 \stackrel{\cdot}{\longrightarrow} F_3$ is defined by $(\sigma\!\cdot\!\tau)A = \sigma A\cdot\tau A$. That this is a natural transformation is seen from

$$
\begin{array}{ccc}
A & \qquad & \begin{array}{ccccc} F_1A & \xrightarrow{\tau A} & F_2A & \xrightarrow{\sigma A} & F_3A \\ {\scriptstyle F_1f}\downarrow & & \downarrow{\scriptstyle F_2f} & & \downarrow{\scriptstyle F_3f} \\ F_1B & \xrightarrow[\tau B]{} & F_2B & \xrightarrow[\sigma B]{} & F_3B \end{array} \\
{\scriptstyle f}\downarrow & & \\
B & &
\end{array}
$$

$\sigma\!\cdot\!\tau$ is called the **horizontal composition** of τ, σ. It is clear that $\mathbf{L}^{\mathbf{K}}$ is a category with identity transformations $\mathrm{id}_F:F \stackrel{\cdot}{\longrightarrow} F$, $(\mathrm{id}_F)A = \mathrm{id}_{FA}$.

2 **DEFINITION:** Let $F:\mathbf{K}\longrightarrow\mathbf{L}$, $G_1, G_2:\mathbf{L}\longrightarrow\mathbf{M}$, $H:\mathbf{M}\longrightarrow\mathbf{N}$ be functors and let $\tau:G_1 \stackrel{\cdot}{\longrightarrow} G_2$ be a natural transformation. Define $\tau F:G_1F \stackrel{\cdot}{\longrightarrow} G_2F$ by $(\tau F)K = \tau(FK)$. Naturality of τF is an instance of the naturality of τ:

$$
\begin{array}{ccccc}
K & \qquad & FK & \qquad & \begin{array}{ccc} G_1FK & \xrightarrow{\tau FK} & G_2FK \\ {\scriptstyle G_1Ff}\downarrow & & \downarrow{\scriptstyle G_2Ff} \\ G_1FK' & \xrightarrow[\tau FK']{} & G_2FK' \end{array} \\
{\scriptstyle f}\downarrow & & {\scriptstyle Ff}\downarrow & & \\
K' & & FK' & &
\end{array}
$$

Define $H\tau:HG_1 \stackrel{\cdot}{\longrightarrow} HG_2$ by $(H\tau)L = H(\tau L)$. Naturality of $H\tau$ follows from the fact that the functor H preserves the appropriate τ-naturality square:

$$
\begin{array}{ccccc}
L & \qquad & \begin{array}{ccc} G_1L & \xrightarrow{\tau L} & G_2L \\ {\scriptstyle G_1f}\downarrow & & \downarrow{\scriptstyle G_2f} \\ G_1L' & \xrightarrow[\tau L']{} & G_2L' \end{array} & \qquad & \begin{array}{ccc} HG_1L & \xrightarrow{H\tau L} & HG_2L \\ {\scriptstyle HG_1f}\downarrow & & \downarrow{\scriptstyle HG_2f} \\ HG_1L' & \xrightarrow[H\tau L']{} & HG_2L' \end{array} \\
{\scriptstyle f}\downarrow & & & & \\
L' & & & &
\end{array}
$$

3 **DEFINITION:** Let $\tau:F \stackrel{\cdot}{\longrightarrow} F' \in \mathbf{L}^{\mathbf{K}}$ and let $\sigma:G \stackrel{\cdot}{\longrightarrow} G' \in \mathbf{M}^{\mathbf{L}}$. The **vertical composition** $\sigma\tau:GF\longrightarrow G'F'$ is defined as the horizontal composition $\sigma\tau = \sigma F'\!\cdot\! G\tau$ of the natural transformations $\sigma F'$ and $G\tau$, and hence is a natural transformation. As we see from the σ-naturality square applied to $\tau K:FK\longrightarrow F'K$

$$
\begin{array}{ccc}
FK & \qquad & \begin{array}{ccc} GFK & \xrightarrow{\sigma FK} & G'FK \\ {\scriptstyle G\tau K}\downarrow & \searrow^{\sigma\tau K} & \downarrow{\scriptstyle G'\tau K} \\ GF'K & \xrightarrow[\sigma F'K]{} & G'F'K \end{array} \\
{\scriptstyle \tau K}\downarrow & & \\
F'K & &
\end{array}
$$

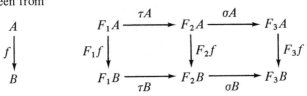

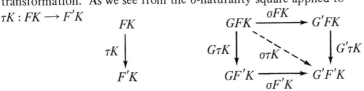

$\sigma\tau$ is also the horizontal composition $G'\tau \cdot \sigma F$.

The following, which appeared in [R. Godement, *Theorie des Faisceaux*, Paris, Hermann, 1958, appendix], are known as **Godement's rules.**

4 **PROPOSITION**: The following identities hold:

GR1. Given $F:\mathbf{K}\longrightarrow\mathbf{L}$, $G:\mathbf{L}\longrightarrow\mathbf{M}$, $H_1, H_2:\mathbf{M}\longrightarrow\mathbf{N}$, $\tau:H_1 \dashrightarrow H_2$,

$$\tau(GF) = (\tau G)F.$$

GR2. Given $F_1, F_2:\mathbf{K}\longrightarrow\mathbf{L}$, $G:\mathbf{L}\longrightarrow\mathbf{M}$, $H:\mathbf{M}\longrightarrow\mathbf{N}$, $\tau:F_1 \dashrightarrow F_2$,

$$(HG)\tau = H(G\tau).$$

GR3. Given $F:\mathbf{K}\longrightarrow\mathbf{L}$, $G_1, G_2:\mathbf{L}\longrightarrow\mathbf{M}$, $H:\mathbf{M}\longrightarrow\mathbf{N}$, $\tau:G_1 \dashrightarrow G_2$,

$$H(\tau F) = (H\tau)F.$$

GR4. Given $F:\mathbf{K}\longrightarrow\mathbf{L}$, $G_1, G_2, G_3:\mathbf{L}\longrightarrow\mathbf{M}$, $H:\mathbf{M}\longrightarrow\mathbf{N}$, $\tau:G_1 \dashrightarrow G_2$, $\sigma:G_2 \dashrightarrow G_3$ then (using notation justified by GR3)

$$H(\sigma\cdot\tau)F = H\sigma F\cdot H\tau F.$$

Proof: The reader can easily supply the requisite diagrams. □

5 **THEOREM**: Let $\mathbf{K}$ be any category. Define

$$\otimes:\mathbf{K}^{\mathbf{K}}\times\mathbf{K}^{\mathbf{K}}\longrightarrow\mathbf{K}^{\mathbf{K}}$$

by $F\otimes G = GF$ on objects and $\tau\otimes\sigma = \sigma\tau$ (vertical composition) on morphisms. Let I be the identity functor of $\mathbf{K}$. Then $(\mathbf{K}^{\mathbf{K}}, \otimes, I, \text{id}, \text{id}, \text{id})$ is a strictly monoidal category.

Proof: (i) Let $\tau_1:F\dashrightarrow F'$, $\tau_2:G\dashrightarrow G'$, $\tau_3:H\dashrightarrow H' \in \mathbf{K}^{\mathbf{K}}$. We then have the diagrams below.

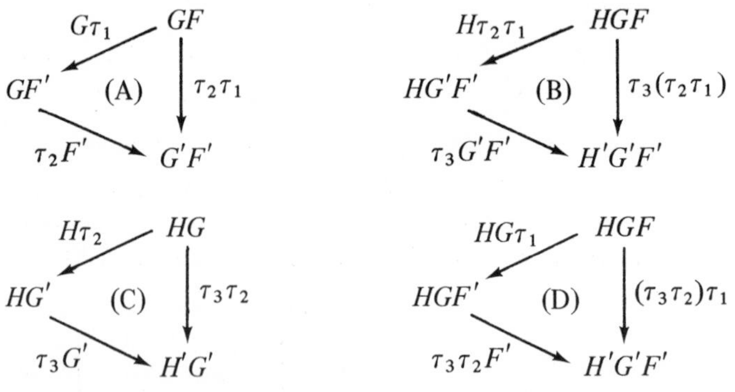

Associativity is then a formal consequence of the Godement rules:

$$\begin{aligned}
\tau_3(\tau_2\tau_1) &= \tau_3 G'F' \cdot H\tau_2\tau_1 && \text{by (B)}\\
&= \tau_3 G'F' \cdot H(\tau_2 F' \cdot G\tau_1) && \text{by (A)}\\
&= \tau_3 G'F' \cdot H\tau_2 F' \cdot HG\tau_1 && \text{by GR4}\\
&= (\tau_3\tau_2)F' \cdot HG\tau_1 && \text{by GR4 and (C)}\\
&= (\tau_3\tau_2)\tau_1 && \text{by (D).}
\end{aligned}$$

(ii) It is also necessary to prove the unitary laws for I and that $\otimes$ is a functor, i.e.

(a) Given $\tau : F \longrightarrow F'$, then $\mathrm{id}_I F = F : IF \longrightarrow IF'$ and $F\mathrm{id}_I = F : FI \longrightarrow FI$.

(b) Given

$$F \xrightarrow{\tau_1} F' \xrightarrow{\tau_2} F''$$

$$G \xrightarrow{\sigma_1} G' \xrightarrow{\sigma_2} G''$$

then $(\sigma_2 \cdot \sigma_1)(\tau_2 \cdot \tau_1) = (\sigma_2\tau_2) \cdot (\sigma_1\tau_1)$.

(c) Given $F, G : \mathbf{K} \longrightarrow \mathbf{K}$, $(\mathrm{id}_F)(\mathrm{id}_G) = \mathrm{id}_{FG}$.

This is left as exercise 2. □

As an easy reworking of the above proof, the reader may then verify:

6 **THEOREM**: Let **Cat** be the category of categories and functors, considered as a monoidal category as in 2.3. Then **Cat** is a **Cat**-category, with respect to the data

$$\mathbf{Cat}[\mathbf{K}, \mathbf{L}] = \mathbf{L}^{\mathbf{K}}$$

$$\mathbf{M}^{\mathbf{L}} \times \mathbf{L}^{\mathbf{K}} \xrightarrow{c_{\mathbf{K},\mathbf{L},\mathbf{M}}} \mathbf{M}^{\mathbf{K}}, \quad G, F \mapsto GF$$

$$1 \xrightarrow{e_{\mathbf{K}}} \mathbf{Cat}[\mathbf{K}, \mathbf{K}] = \mathrm{id}_{\mathbf{K}} : \mathbf{K} \longrightarrow \mathbf{K}. \qquad \square$$

Exercises

1 Prove Proposition 4.

2 Complete the proof of Theorem 5.

3 Explore the structure of the monoidal category $\mathbf{K}^{\mathbf{K}}$ when **K** has one object. [Hint: If **K** is the monoid (X, m, e), an object is a monoid homomorphism from (X, m, e) to itself and a morphism from f to g is an element $\tau \in X$ satisfying $m(\tau, f(x)) = m(g(x), \tau)$; e.g., in a group, τ is a morphism from $f(x)$ to $\tau f(x)\tau^{-1}$.]

4 Explore the structure of the monoidal category $\mathbf{K}^{\mathbf{K}}$ when **K** is a poset.

5 Prove that $(\mathbf{L}^{\mathrm{op}})^{\mathbf{K}^{\mathrm{op}}}$ is isomorphic to $\mathbf{L}^{\mathbf{K}}$.

6 Prove that if **L** has small limits (exercise 2.4.13) then so, too, does $\mathbf{L}^{\mathbf{K}}$ for any **K**. [Hint: If D is a diagram in $\mathbf{L}^{\mathbf{K}}$ and if $E(K)$ is the limit of $D_i(K)$, $E \in \mathbf{L}^{\mathbf{K}}$ is the desired limit.]

7 Let **K** be a **small category**, i.e. the objects of **K** as well as the morphisms of **K** constitute a set. Prove that $\mathbf{Set}^{\mathbf{K}}$ is a topos. [Hints: 6 provides products and a terminal object. To prove that $-\times F: \mathbf{Set}^{\mathbf{K}} \longrightarrow \mathbf{Set}^{\mathbf{K}}$ has a right adjoint $G \mapsto G^F$ combine the requirement

$$\frac{H \times F \longrightarrow G}{H \longrightarrow G^F}$$

with the Yoneda lemma, with $H = \mathbf{K}(K, -)$, to conclude $G^F(K) = \{\alpha | \alpha : \mathbf{K}(K, -) \times H \dot{\longrightarrow} G\}$. Similarly, $\Omega(K) = \{S \mid S$ is a subfunctor of $\mathbf{K}(K, -)\}$ where **subfunctor** means monosubobject in $\mathbf{Set}^{\mathbf{K}}$ (and it is easy to prove S is a subfunctor of F if and only if $S(K) \subset F(K)$ and inclusion $S \dot{\longrightarrow} F$ is natural).]

Chapter 10

MONADS AND ALGEBRAS

Throughout this book, we have seen how the basic concepts of many different branches of mathematics find unified expression in the language of category theory. Particularly striking is the diversity of mathematical constructions which have proved to be examples of adjunctions. However, while category theory has been evolving its general perspective on mathematics, another general algebraic perspective – called **universal algebra** – has evolved to unify the study of semigroups, monoids, groups, rings, and vector spaces, amongst other algebraic structures. The theory of **monads** (also known as **triples**) brings the insight of universal algebra into the categorical framework, and it is with an introduction to this 'bridge' that we close this book.

10.1 UNIVERSAL ALGEBRA

In this section we provide the basic definitions of universal algebra, so that the reader may see how the theory of monads makes contact with this approach.

1 **DEFINITION**: Let Ω be a set together with† a map $\nu : \Omega \longrightarrow \mathbf{N}$. We say (Ω, ν) is a **label set** and omit mention of ν when no ambiguity can arise. For each $n \in \mathbf{N}$, we call $\Omega_n = \nu^{-1}(n)$ the set of ***n*-ary labels** in Ω. An **Ω-algebra** is then a set Q together with a function

$$\delta_\omega : Q^n \longrightarrow Q$$

for each $\omega \in \Omega_n$, and each $n \in \mathbf{N}$.

It is interesting to recast this in categorical form:

2 **OBSERVATION**: Given (Ω, ν), define the functor $X_\Omega : \mathbf{Set} \longrightarrow \mathbf{Set}$ by requiring it to act on objects by

$$X_\Omega Q = \coprod_{\omega \in \Omega} Q^{\nu(\omega)}$$

---

† In some applications (e.g. to describe Boolean σ-rings) it is necessary to let the domain of ν include infinite ordinals. However, this would obscure the present development.

i.e. a disjoint union with one copy of Q^n for each $n \in \Omega_n$; while for $f: Q \longrightarrow Q'$ we define

$$X_\Omega f: X_\Omega Q \longrightarrow X_\Omega Q' : (q_1, ..., q_n, \omega) \mapsto (fq_1, ..., fq_n, \omega).$$

Then an Ω-algebra is simply a map

$$\delta : X_\Omega Q \longrightarrow Q.$$

Proof: That X_Ω is a functor is a simple exercise. Finally, it is the nature of coproducts that to give $\delta : X_\Omega Q \longrightarrow Q$ is just to give each $\delta_\omega = \delta \cdot \mathrm{in}_\omega : Q^{\nu(\omega)} \longrightarrow Q$. □

We may thus denote an Ω-algebra by the pair (Q, δ). This categorical recasting lets us define Ω-algebra homomorphisms very simply:

3 **DEFINITION**: Given two Ω-algebras (Q, δ) and (Q', δ'), a homomorphism is simply a map $h: Q \longrightarrow Q'$ which satisfies the diagram

$$\begin{array}{ccc} X_\Omega Q & \xrightarrow{\ \delta\ } & Q \\ {\scriptstyle X_\Omega h}\downarrow & & \downarrow{\scriptstyle h} \\ X_\Omega Q' & \xrightarrow[\ \delta'\ ]{} & Q' \end{array}$$

The reader will see that this diagram unpacks to

$$h\delta_\omega(q_1, ..., q_n) = \delta'_\omega(hq_1, ..., hq_n) \quad \text{for all } (q_1, ..., q_n) \in Q^n, \omega \in \Omega_n, n \in \mathbf{N}$$

which is the usual statement in the universal algebra literature. Whether by a straightforward splicing of diagrams or a manipulation of such equations, we immediately deduce:

4 **FACT**: ⟨Ω-algebras and homomorphisms⟩ form a category, denoted **Ω-Alg**. □

Given Ω, we may build composite expressions, or *terms*. Before we turn to a formal definition, let's see how this relates to group theory.

5 **EXAMPLE**: Let $\Omega = \{m, i, e\}$, where $\nu(m) = 2$, $\nu(i) = 1$, $\nu(e) = 0$. Thus an Ω-algebra is a set Q equipped with two functions $\delta_m : Q^2 \longrightarrow Q$ and $\delta_i : Q \longrightarrow Q$ and a constant δ_e in Q. In general, then, such an algebra will *not* be a group. In fact, it will only be a group if the following conditions are satisfied:

Associativity: $\delta_m(q_1, \delta_m(q_2, q_3)) = \delta_m(\delta_m(q_1, q_2), q_3)$

Identity: $\delta_m(\delta_e, q) = q = \delta_m(q, \delta_e)$

Inverse: $\delta_m(q, \delta_i(q)) = \delta_e = \delta_m(\delta_i(q), q)$.

We may express associativity, then, by saying that δ assigns the same value to the 'trees'

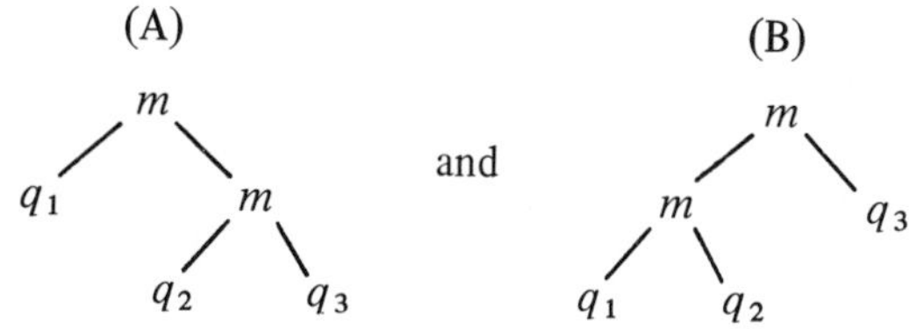

The identity condition requires equal assignments for

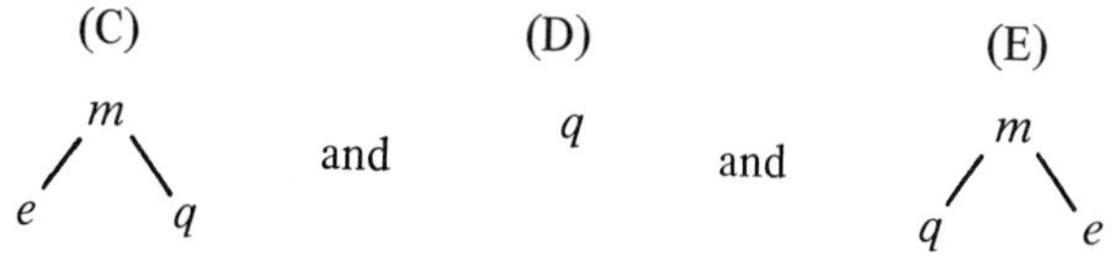

while the inverse condition applies to

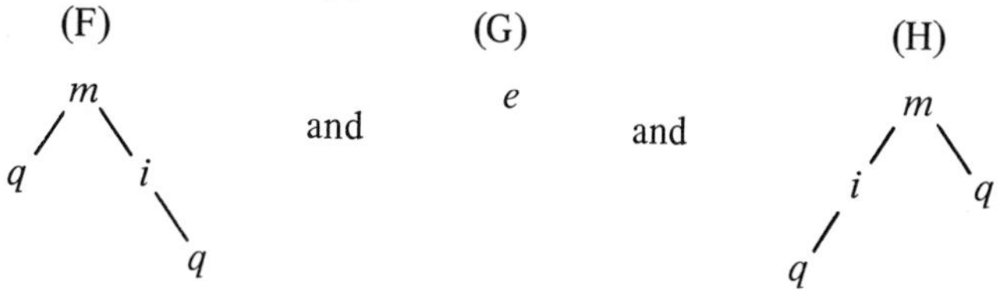

Thus to use the notion of universal algebra to characterize groups, we must be able to talk of Ω-trees, and must be able to restrict attention to δ's which 'run' to identical values on stipulated pairs of trees. We now formalize these notions:

6 **DEFINITION**: Given a set Q and label set (Ω, ν) we define the set of **Ω-trees over** Q to be the set $T_\Omega Q$ defined inductively by:

(a) $T_\Omega Q$ contains Q and Ω_0

(b) If $\omega \in \Omega_n$, and $t_1, \ldots, t_n$ are in $T_\Omega Q$, then $\omega[t_1, \ldots, t_n]$ is also in $T_\Omega Q$.

We thus have trees (D) and (G) of the previous example included in $T_\Omega Q$ by clause (a) – since $q \in Q$ and $e \in \Omega_0$ – while, for example, (B) is represented by the expression

$$m[m[q_1, q_2], q_3] \tag{B'}$$

and $m[q_1, q_2]$ and q_3 are both clearly elements of $T_\Omega Q$. (We won't distinguish linear formulas of the kind (B′) from their tree displays of the kind (B), but will use whichever notation is more convenient.)

7 **OBSERVATION**: We obtain a functor $T_\Omega : \mathbf{Set} \longrightarrow \mathbf{Set}$ when, for each map $f: Q \longrightarrow Q'$ we define $T_\Omega f: T_\Omega Q \longrightarrow T_\Omega Q'$ inductively by

$$T_\Omega f(q) = f(q) \qquad \text{for } q \in Q$$
$$T_\Omega f(\omega) = \omega \qquad \text{for } \omega \in \Omega_0$$
$$T_\Omega f(\omega[t_1, \ldots, t_n]) = \omega[T_\Omega f(t_1), \ldots, T_\Omega f(t_n)].$$

□

Thus $T_\Omega f$ simply relabels each $q \in Q$ leaf with $f(q) \in Q'$, and otherwise leaves the tree unchanged.

We have motivated the construction of T_Ω by the need to equate 'runs' on trees. Let us, then, define such runs and then show, suprisingly, that T_Ω can be obtained by an adjunction.

8 **DEFINITION**: Let (Q, δ) be an Ω-algebra. Then we define $\delta^* : T_\Omega Q \longrightarrow Q$ inductively by the definition:

$$\delta^*(q) = q \qquad \text{for each } q \in Q$$

$$\delta^*(\omega) = \delta_\omega \qquad \text{for each } \omega \in \Omega_0$$

$$\delta^*(\omega[t_1, \ldots, t_n]) = \delta_\omega(\delta^*(t_1), \ldots, \delta^*(t_n)) \qquad \text{for each } \omega \in \Omega_n,\ t_1, \ldots, t_n \in T_\Omega Q.$$

The reader should check that this agrees with the intuition of Example 5, of starting at the leaves of a tree t, and using δ to work up towards the 'root', at which the final value $\delta^*(t)$ is obtained.

9 **DEFINITION**: Let (Ω, ν) be a label set. Then a set E of **Ω-equations** is a subset of $T_\Omega I \times T_\Omega I$ for some fixed set I.

10 **EXAMPLE**: We may recapture the identities of Example 5 by taking $I = \{1, 2, 3\}$ and by using the set E of five equations:

$$(m[1, m[2, 3]], m[m[1, 2], 3])$$

$$(m[e, 1], 1) \quad \text{and} \quad (1, m[1, e])$$

$$(m[1, i[1]], e) \quad \text{and} \quad (e, m[i[1], 1]).$$

We then have for this Ω and E that an Ω-algebra (Q, δ) is a group iff for every map $f: I \longrightarrow Q$ and for every equation (t_1, t_2) in E we have

$$\delta^*[T_\Omega f(t_1)] = \delta^*[T_\Omega f(t_2)].$$

For example, for $(t_1, t_2) = (m[e, 1], 1)$ and $f(1) = q$, we have

$$T_\Omega f(t_1) = (m[e, q]) \quad \text{and} \quad T_\Omega f(t_2) = q$$

and applying δ^* yields the desired equality

$$\delta_m(\delta_e, q) = q.$$

Thus groups are a special case of the following:

11 **DEFINITION**: Let (Ω, ν) be a label set and $E \subset T_\Omega I \times T_\Omega I$ a set of Ω-equations. Then an **(Ω, E)-algebra** is an Ω-algebra which **satisfies** E, i.e. for every $(t_1, t_2) \in E$ and every $f: I \longrightarrow Q$, we have

$$\delta^*[T_\Omega f(t_1)] = \delta^*[T_\Omega f(t_2)].$$

(Ω, E)-**Alg** is the full subcategory of Ω-**Alg** whose objects are the (Ω, E)-algebras.

12 **THEOREM**: The forgetful functor Ω-**Alg** $\longrightarrow$ **Set** has a left adjoint. The free Ω-algebra over I has the form $(T_\Omega I, \mu_0 I)$.

Proof: Contemplate the diagrams

$$\begin{array}{ccc} I & \xrightarrow{\eta I} & T_\Omega I \\ & {}_{f}\searrow \quad (A) & \downarrow \psi \\ & & Q \end{array} \qquad\qquad \begin{array}{ccc} X_\Omega T_\Omega I & \xrightarrow{\mu_0 I} & T_\Omega I \\ {\scriptstyle X_\Omega \psi}\downarrow & (B) & \downarrow {\scriptstyle \psi} \\ X_\Omega Q & \xrightarrow[\delta]{} & Q \end{array}$$

$\eta I : I \longrightarrow T_\Omega I$ is to send an element $i \in I$ to an Ω-tree. The obvious choice is $\eta I(i) = i$, enabled by clause (a) of 6. Again the obvious choice for $\mu_0 I$ is

$$X_\Omega T_\Omega I \longrightarrow T_\Omega I : (t_1, \ldots, t_n, \omega) \mapsto \omega[t_1, \ldots, t_n]$$

which works by clause (b) of 6. But then, given $f : I \longrightarrow Q$, (A) just says

$$\psi(i) = f(i) \qquad \text{for each } i \in I$$

while (B) provides the induction step

$$\psi(\omega[t_1, \ldots, t_n]) = \delta_\omega(\psi(t_1), \ldots, \psi(t_n))$$

which includes $\psi(\omega) = \delta_\omega$ for the case $\omega \in \Omega_0$. This defines ψ uniquely. In fact, we see from Definition 8 that

$$\psi(t) = \delta^*[T_\Omega f(t)].$$

Thus the pair $((T_\Omega I, \mu_0 I), \eta_I)$ is free over I with respect to U. □

Exercises

1 Verify that X_Ω is a functor **Set** $\longrightarrow$ **Set** and that Ω-**Alg** is indeed a category of sets with structure.

2 In Section 6.3 we studied sequential machines and noted that their dynamics formed a category **Dyn**(X_0). Generalizing from the functor $-\times X_0 :$ **Set** $\longrightarrow$ **Set** to an arbitrary functor $X : \mathbf{K} \longrightarrow \mathbf{K}$, define a category **Dyn**(X) whose objects are X-**dynamics** (Q, δ) – i.e., **K**-morphisms $\delta : XQ \longrightarrow Q$ – and whose morphisms $h : (Q, \delta) \longrightarrow (Q', \delta')$ are **K**-morphisms $h : Q \longrightarrow Q'$ satisfying

$$\begin{array}{ccc} XQ & \xrightarrow{\delta} & Q \\ {\scriptstyle Xh}\downarrow & & \downarrow {\scriptstyle h} \\ XQ' & \xrightarrow{\delta'} & Q' \end{array}$$

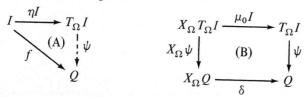

(a) Verify that **Dyn**(X) is a category.

(b) Verify that **Dyn**(X_Ω) = Ω-**Alg**.

(c) Given a family (Q_i, δ_i) of X-dynamics and a product diagram $Q \xrightarrow{\pi_i} Q_i$ in **K** prove that $\pi_i : (Q, \delta) \longrightarrow (Q_i, \delta_i)$ is a product diagram in **Dyn**(X) where δ is the unique morphism indicated below:

$$\begin{array}{ccc} XQ & \overset{\delta}{\dashrightarrow} & Q \\ {\scriptstyle X\pi_i}\downarrow & & \downarrow{\scriptstyle \pi_i} \\ XQ_i & \underset{\delta_i}{\longrightarrow} & Q_i \end{array}$$

(d) Similarly, given $f, g : (Q_1, \delta_1) \longrightarrow (Q_2, \delta_2)$ in **Dyn**(X) and $h : Q \longrightarrow Q_1 = \mathrm{eq}(f, g)$ in **K**, show that there exists unique $\delta : XQ \longrightarrow Q$ such that $h : (Q, \delta) \longrightarrow (Q_1, \delta_1)$ is a morphism in **Dyn**(X) and that, moreover, $h = \mathrm{eq}(f, g)$ in **Dyn**(X).

(e) Conclude that if X: **Set** $\longrightarrow$ **Set**, then the underlying set functor **Dyn**(X) $\longrightarrow$ **Set** has a left adjoint if and only if it satisfies the solution set condition.

3 Present each of **Mon** and **Grp** as a category of (Ω, E)-algebras.

4 Present **Vect** as a category of (Ω, E)-algebras. [Hint: Use a unary label for each scalar.]

5 Show that neither **Met** nor **Top** can be presented as a category of (Ω, E)-algebras. [Hint: Consider 2c, 2d above.] In more precise terms, we are asserting that neither **Met** nor **Top** is isomorphic to (Ω, E)-**Alg** as a category of sets with structures; it is conceivable that there is a perverse isomorphism **Met** $\longrightarrow$ (Ω, E)-**Alg**, but it cannot commute with the underlying set functors.

6 Define X: **Set** $\longrightarrow$ **Set** by

$$XQ = \{A \mid A \text{ is a non-empty subset of } Q\},$$

and, for $f : Q \longrightarrow Q'$, $(Xf)(A) = \{fa \mid a \in A\}$.
Verify that X is a functor. Show that X cannot have the form X_Ω for any Ω. [Hint: Consideration of Q with at most two elements leads to the contradiction "$3 = 2^n$ for some n".] Prove that **Dyn**(X) $\longrightarrow$ **Set** has a left adjoint. [Hint: See exercise 2.]

7 (J. R. Isbell) Construct an (Ω, E) such that $\emptyset$ and 1 are the only finite (Ω, E)-algebras but such that infinite (Ω, E)-algebras exist. [Hint: Use one binary operation m and two unary operations u_1, u_2 and equations asserting that m is bijective.]

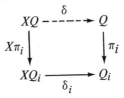

10.2 FROM MONOIDS TO MONADS

We first see how a monoid in **Set** may be re-viewed as a monoid in the functor category $\mathbf{Set}^{\mathbf{Set}}$; and then introduce **monads** as monoids in $\mathbf{K}^{\mathbf{K}}$. Finally, we associate algebras with monads.

If $f: A \longrightarrow B$ is any function then

$$\begin{array}{ccc} Q \times A & \xrightarrow{g \times \mathrm{id}} & Q' \times A \\ {\scriptstyle \mathrm{id} \times f}\downarrow & & \downarrow{\scriptstyle \mathrm{id} \times f} \\ Q \times B & \xrightarrow[g \times \mathrm{id}]{} & Q' \times B \end{array}$$

commutes for every $g: Q \longrightarrow Q'$. In other words, $\mathrm{id} \times f: - \times A \dot{\longrightarrow} - \times B$ is a natural transformation of functors $\mathbf{Set} \longrightarrow \mathbf{Set}$. Conversely

1 **LEMMA**: If A, B are sets and $\tau: - \times A \dot{\longrightarrow} - \times B$ is a natural transformation there exists unique $f: A \longrightarrow B$ such that $\tau Q = \mathrm{id}_Q \times f$ for all sets Q.

Proof: Define f by

$$f: A \xrightarrow{\lambda_A^{-1}} 1 \times A \xrightarrow{\tau 1} 1 \times B \xrightarrow{\lambda_B} B$$

where λ_A, λ_B are the projections. Let Q be any set and let $q: 1 \longrightarrow Q$ be an element of Q.

$$\begin{array}{ccccccc} A & \xrightarrow{\lambda_A^{-1}} & 1 \times A & \xrightarrow{\tau 1} & 1 \times B & \xrightarrow{\lambda_B} & B \\ & & {\scriptstyle q \times \mathrm{id}}\downarrow & & \downarrow{\scriptstyle q \times \mathrm{id}} & \nearrow{\scriptstyle \pi_2} & \\ & & Q \times A & \xrightarrow[\tau Q]{} & Q \times B & & \end{array}$$

Then for any $a \in A$ we have $(\tau Q)(q, a) = (\tau Q)(q \times \mathrm{id})(\lambda_A^{-1})(a) =$ (using the naturality square above) $(q \times \mathrm{id})(\tau 1)(\lambda_A^{-1})(a) = (q, \pi_2(q \times \mathrm{id})(\tau 1)(\lambda_A^{-1})(a)) =$ $(q, f(a))$ as desired. The uniqueness of f is clear looking only at $Q = 1$. □

Lemma 1 allows us to say "sets are functors." We associate to each set A the functor $- \times A : \mathbf{Set} \longrightarrow \mathbf{Set}$. By evaluating at 1, this passage is injective. Functions $f: A \longrightarrow B$ are in bijective correspondence with natural transformations $\mathrm{id} \times f: - \times A \dot{\longrightarrow} - \times B$ by the lemma. Moreover $\mathrm{id} \times f$ is the identity transformation if f is an identity function and, given $g: B \longrightarrow C$, $\mathrm{id} \times (gf) = (\mathrm{id} \times g) \cdot (\mathrm{id} \times f)$. In short, **Set** may be regarded as a full subcategory of $\mathbf{Set}^{\mathbf{Set}}$. **Set** and $\mathbf{Set}^{\mathbf{Set}}$ are both monoidal categories (9.2.3, 9.4.5) and the monoidal structures are closely related. For observe that the composition $(- \times B) \cdot (- \times A)$

evaluated at Q gives $(Q \times A) \times B$ whereas $- \times (A \times B)$ evaluates as $Q \times (A \times B)$. Hence, up to isomorphism, $\times : \mathbf{Set} \times \mathbf{Set} \longrightarrow \mathbf{Set}$ is represented by composition in $\mathbf{Set}^{\mathbf{Set}}$. Thus an ordinary monoid (X, m, e) can be described as a monoid in the monoidal category $\mathbf{Set}^{\mathbf{Set}}$ as follows. Let $\tilde{X}: \mathbf{Set} \longrightarrow \mathbf{Set}$ be the functor $- \times X$, and write $\tilde{m} = \mathrm{id} \times m : \tilde{X}\tilde{X} \dot{\longrightarrow} \tilde{X}$, i.e.

$$Q\tilde{m} = (Q \times X) \times X \cong Q \times (X \times X) \xrightarrow{\mathrm{id}_Q \times m} Q \times X .$$

Then $\tilde{m}$ is a natural transformation satisfying the (strict) associative law

$$\begin{array}{ccc} \tilde{X}\tilde{X}\tilde{X} & \xrightarrow{\tilde{X}\tilde{m}} & \tilde{X}\tilde{X} \\ {\scriptstyle \tilde{m}\tilde{X}}\downarrow & & \downarrow{\scriptstyle \tilde{m}} \\ \tilde{X}\tilde{X} & \xrightarrow[\tilde{m}]{} & \tilde{X} \end{array}$$

Let $\tilde{e}: I_{\mathbf{Set}} \dot{\longrightarrow} \tilde{X}$ be the natural transformation $\tilde{e}A : A \longrightarrow A \times X,\ a \mapsto (a, e)$. Then $(\tilde{m}, \tilde{e})$ satisfy the unitary laws

$$\begin{array}{ccccc} \tilde{X} & \xrightarrow{\tilde{X}\tilde{e}} & \tilde{X}\tilde{X} & \xleftarrow{\tilde{e}\tilde{X}} & \tilde{X} \\ & {\scriptstyle \mathrm{id}}\searrow & \downarrow{\scriptstyle \tilde{m}} & \swarrow{\scriptstyle \mathrm{id}} & \\ & & \tilde{X} & & \end{array}$$

Indeed it is routine to show:

2 **OBSERVATION**: Let X be a set, and let $m : X \times X \longrightarrow X$, $e : 1 \longrightarrow X$ giving rise to the functor $\tilde{X}$ and natural transformations $\tilde{m} : \tilde{X}\tilde{X} \dot{\longrightarrow} \tilde{X}$ and $\tilde{e} : I_{\mathbf{Set}} \dot{\longrightarrow} \tilde{X}$. Then (X, m, e) is an ordinary monoid in **Set** iff $(\tilde{X}, \tilde{m}, \tilde{e})$ is a monoid in $\mathbf{Set}^{\mathbf{Set}}$. □

In some sense, then, monoids can be defined in *any* category:

3 **DEFINITION**: Let **K** be an arbitrary category and let $\mathbf{K}^{\mathbf{K}}$ be the strict monoidal category of 9.4.5. A **monad in K** is a monoid in $\mathbf{K}^{\mathbf{K}}$. Thus, a monad in **K** is a triple $\mathbf{T} = (T, \eta, \mu)$ where $T: \mathbf{K} \longrightarrow \mathbf{K}$ is a functor, $\eta : \mathrm{id}_{\mathbf{K}} \dot{\longrightarrow} T$ and $\mu : TT \dot{\longrightarrow} T$ are natural transformations subject to the **T-associative law**

$$\begin{array}{ccc} TTT & \xrightarrow{T\mu} & TT \\ {\scriptstyle \mu T}\downarrow & & \downarrow{\scriptstyle \mu} \\ TT & \xrightarrow[\mu]{} & T \end{array}$$

and the **T-unitary laws**

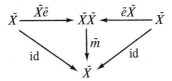

$$\begin{array}{ccccc} T & \xrightarrow{T\eta} & TT & \xleftarrow{\eta T} & T \\ & {\scriptstyle \text{id}}\searrow & \downarrow{\scriptstyle \mu} & \swarrow{\scriptstyle \text{id}} & \\ & & T & & \end{array}$$

It is not hard to find examples of monads:

4 THEOREM: Let $G: \mathbf{A} \longrightarrow \mathbf{K}$ have a left adjoint $F: \mathbf{K} \longrightarrow \mathbf{A}$ giving rise to the natural transformations $\eta: I_{\mathbf{K}} \dot{\longrightarrow} T = GF$ and $\varepsilon: FG \dot{\longrightarrow} I_{\mathbf{A}}$ as in 7.3.9. Then $\mathbf{T} = (T, \eta, \mu)$, where $\mu: TT \dot{\longrightarrow} T = G\varepsilon F: G(FG)F \dot{\longrightarrow} GF$, is a monad in $\mathbf{K}$.

Proof: By construction, T is a functor and η, μ are natural transformations. The $\mathbf{T}$-associative law is

$$\begin{array}{ccc} GFGFGF & \xrightarrow{GFG\varepsilon F} & GFGF \\ {\scriptstyle G\varepsilon FGF}\downarrow & & \downarrow{\scriptstyle G\varepsilon F} \\ GFGF & \xrightarrow[G\varepsilon F]{} & GF \end{array}$$

which is simply the image under F for the ε-naturality square for the morphism $G\varepsilon$. The $\mathbf{T}$-unitary laws follow from the identities of (10) of 7.3.9

$$\begin{array}{ccccc} GF & \xrightarrow{GF\eta} & GFGF & \xleftarrow{\eta GF} & GF \\ & {\scriptstyle \text{id}}\searrow & \downarrow{\scriptstyle G\varepsilon F} & \swarrow{\scriptstyle \text{id}} & \\ & & GF & & \end{array}$$

i.e. $G\varepsilon F \cdot GF\eta = G(\varepsilon F \cdot F\eta) = G(\text{id}_F) = \text{id}_{GF}$ and $G\varepsilon F \cdot \eta GF = (G\varepsilon \cdot \eta G)F$ $= (\text{id}_G)F = \text{id}_{GF}$. □

In particular, we may tie this back to Theorem 1.12.

5 DEFINITION: We denote by $\mathbf{T}_\Omega = (T_\Omega, \eta_\Omega, \mu_\Omega)$ the monad associated with the forgetful functor U: Ω-**Alg** $\longrightarrow$ **Set**. We sometimes call $\mathbf{T}_\Omega$ the (Ω-) **tree monad**.

We saw that $\eta_Q : I \dot{\longrightarrow} T_\Omega$ is the inclusion of generators defined on elements by $\eta_Q(i) = i$. Now $T_\Omega T_\Omega I$ is the set of Ω-trees on generators which are Ω-trees on I generators. $\mu_\Omega I: T_\Omega T_\Omega I \longrightarrow T_\Omega I$ is then simply 'unfolding' so that, for example, if

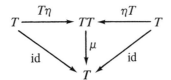

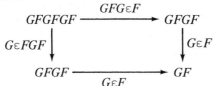

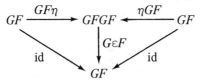

t is 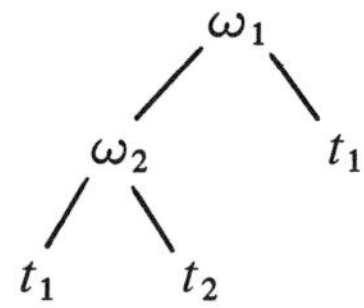 ; where t_1 is 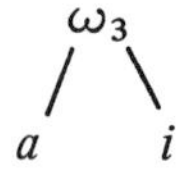 and t_2 is c

then $\mu_\Omega I(t)$ is 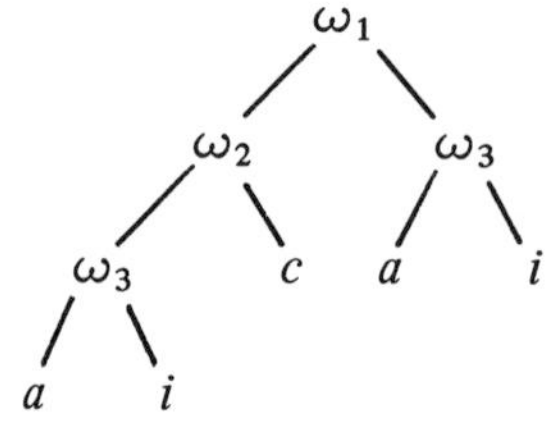

Let us now see how the notion of "algebras over a monad" generalizes "actions of a monoid": Let (X, m, e) be an ordinary monoid. An *action of X on a* set Q is a monoid homomorphism $\psi : X \longrightarrow Q^Q : x \mapsto \psi_x$ from X to the monoid of functions from Q to Q under functional composition. Under the transposition

$$\frac{X \xrightarrow{\psi} Q^Q}{Q \times X \xrightarrow{\xi} Q}$$

resulting from the cartesian closed structure of **Set** (i.e., define $\xi(q, x) = \psi_x(q)$) the condition that ψ be a monoid homomorphism may be expressed as a condition on ξ:

6 **DEFINITION**: An **action** of the monoid (X, m, e) on the set Q is a function $\xi : Q \times X \longrightarrow Q : (q, x) \mapsto q \cdot x$ satisfying the laws

$$q \cdot (x \cdot x') = (q \cdot x) \cdot x'$$
$$q \cdot e = q.$$

There are many examples of actions. For example, let (Q, d) be a metric space and let G be the geometry of (Q, d) as in exercise 4 of Section 4.1. As G is a submonoid (indeed a subgroup) of Q^Q the inclusion map $\psi : G \longrightarrow Q^Q$ is a monoid homomorphism. The corresponding $\xi : Q \times G \longrightarrow Q$ is given by $\xi(q, g) = g(q)$; this formulation seems appropriate in that "rigid motions act on points."

7 **EXAMPLE**: In our discussion of automata theory, we saw how to pass from a dynamics $\delta : Q \times X_0 \longrightarrow Q$ to its run map $\delta^* : Q \times {X_0}^* \longrightarrow Q$. We defined $\delta^*(q, \Lambda)$ to be q for any q in Q, and the induction step

$$\delta^*(q, wx) = \delta(\delta^*(q, w), x)$$

easily yields that $\delta^*(q, ww') = \delta^*(\delta^*(q, w), w')$ for all q in Q and w, w' in $X_0{}^*$. It is thus clear that there is a bijection between X_0-dynamics and actions of the free monoid $X_0{}^*$.

Note that tree monads include free monoids: Let $\Omega_1 = X_0$, and Ω_n be empty for $n \neq 1$. Then for each set I, $T_\Omega I \cong I \times X_0{}^*$, and we have that

$$\mu I : TTI \longrightarrow TI : (I \times X_0{}^*) \times X_0{}^* \longrightarrow I \times X_0{}^* : ((i, w), w') \mapsto (i, ww')$$

which the reader will recognize as the *run map of the free dynamics* $\mu_0 I : (I \times X_0{}^*) \times X_0 \longrightarrow I \times X_0{}^* : ((i, w), x) \mapsto (i, wx)$ *on I generators.*

8 **OBSERVATION**: Let (X, m, e) be a monoid and let $\xi : Q \times X \longrightarrow Q$ be a function. Let $(\tilde{X}, \tilde{m}, \tilde{e})$ be the corresponding monad in **Set** as in 2. Then (Q, ξ) is an action of X iff we have

$$\begin{array}{ccc} Q & \xrightarrow{\tilde{e}Q} & \tilde{X}Q \\ & {\scriptstyle \mathrm{id}} \searrow & \downarrow{\scriptstyle \xi} \\ & & Q \end{array} \qquad\qquad \begin{array}{ccc} \tilde{X}\tilde{X}Q & \xrightarrow{\tilde{X}\xi} & \tilde{X}Q \\ {\scriptstyle \tilde{m}Q}\downarrow & & \downarrow{\scriptstyle \xi} \\ \tilde{X}Q & \xrightarrow[\xi]{} & Q \end{array}$$

□

This motivates the following general definition.

9 **DEFINITION**: Let $\mathbf{T} = (T, \eta, \mu)$ be a monad in **K**. A **T-algebra** is a pair (Q, ξ) where $\xi : TQ \longrightarrow Q$ satisfies the **algebra laws**

$$\begin{array}{ccc} Q & \xrightarrow{\eta Q} & TQ \\ & {\scriptstyle \mathrm{id}_Q} \searrow & \downarrow{\scriptstyle \xi} \\ & & Q \end{array} \qquad\qquad \begin{array}{ccc} TTQ & \xrightarrow{T\xi} & TQ \\ {\scriptstyle \mu Q}\downarrow & & \downarrow{\scriptstyle \xi} \\ TQ & \xrightarrow[\xi]{} & Q \end{array}$$

Further discussion of these axioms will be provided in the next section.

Exercises

1 Let $T : \mathbf{Set} \longrightarrow \mathbf{Set}$ be the identity functor. Show that $\mathbf{T} = (T, \mathrm{id}, \mathrm{id})$ is a monad in **Set** and that each set possesses a unique **T**-algebra structure.

2 Define $T : \mathbf{Set} \longrightarrow \mathbf{Set}$ by $TQ = Q + 1$,

$$\begin{array}{c} Q \\ {\scriptstyle f}\downarrow \\ Q' \end{array} \qquad\qquad \begin{array}{ccccc} Q & \xrightarrow{\mathrm{in}_1} & Q+1 & \xleftarrow{\mathrm{in}_2} & 1 \\ {\scriptstyle f}\downarrow & & \downarrow{\scriptstyle Tf} & \swarrow{\scriptstyle \mathrm{in}_2} & \\ Q' & \xrightarrow[\mathrm{in}_1]{} & Q'+1 & & \end{array}$$

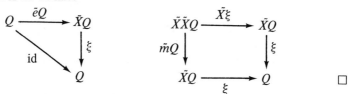

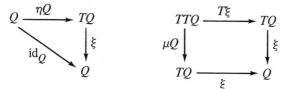

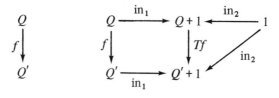

Set $\eta Q = \text{in}_1$, and define μ by

$$Q\mu = (Q + 1) + 1 \cong Q + (1 + 1) \xrightarrow{\text{id} + z} Q + 1$$

where $z : 1 + 1 \longrightarrow 1$ is the unique function. Show that $\mathbf{T} = (T, \eta, \mu)$ is a monad in **Set**. What are the **T**-algebras? Rework the entire construction in the category of metric spaces of diameter at most 1.

3 Define T: **Set** $\longrightarrow$ **Set** by $TQ = \{A \mid A \text{ is a subset of } Q\}$, $(Tf)(A) = \{fa \mid a \in A\}$. Define $(eQ)(q) = \{q\}$ and, for $\mathbf{A}$ in $T(TQ)$, $(\mu Q)(\mathbf{A}) = \cup \mathbf{A} = \{q \in Q \mid q \in A \text{ for some } A \in \mathbf{A}\}$. Show that $\mathbf{T} = (T, \eta, \mu)$ is a monad in **Set**. Show that for each set Q, the passage from poset structures $\leqslant$ such that every subset of Q has a supremum in $(Q, \leqslant)$ to **T**-algebra structures $\xi : TQ \longrightarrow Q$, defined by $\xi(A) = \sup(A)$, is well-defined and bijective. [Hint: To go the other way define $q \leqslant q'$ if $\xi\{q, q'\} = q'$.]

4 Let (X, d) be a metric space. Given a sequence (x_n) in X and an element x of X we say (x_n) **converges to** x if for every $\varepsilon > 0$ (no matter how small) there exists N (perhaps large, depending on ε) such that $d(x_n, x) < \varepsilon$ whenever $n > N$. Given $A \subset X$, the **closure of** A is the subset

$$A^- = \{x \in X \mid \text{there exists } (a_n) \text{ converging to } x \text{ with each } a_n \text{ in } A\}.$$

(a) Verify that $A \subset A^-$ and that $(A^-)^- = A^-$.

(b) Let **K** be the poset-qua-category of all subsets of X ordered by inclusion. Let T: **K** $\longrightarrow$ **K** be defined by $TA = A^-$. Verify that $\mathbf{T} = (T, \eta, \mu)$ is a monad for unique η and μ and that **T**-algebras are coextensive with the **closed** subsets, i.e. those subsets A satisfying $A = A^-$.

5 Let $\mathbf{T} = (T, \eta, \mu)$ be a monad in **K** and let Q be in **K**. Show that $(TQ, \mu Q)$ is a **T**-algebra.

10.3 MONADS FROM FREE ALGEBRAS

We now give a completely different motivation for monads than that of Section 2.

Let us begin with a convenient category of (Ω, E)-algebras, say the category **Mon** of monoids, and its underlying set functor U: **Mon** $\longrightarrow$ **Set**. For each set Q let $(FQ, \eta Q)$ be the free monoid over Q with respect to U, as in 7.2.2. Thus the underlying set $TQ = UFQ$ of FQ is the set Q^*. Now let (Q, m, e) be a monoid. Then there exists

1
$$\begin{array}{ccc} Q & \xrightarrow{\eta Q} & TQ \\ & \searrow^{\mathrm{id}_Q} & \downarrow \xi \\ & & Q \end{array} \qquad \begin{array}{c} FQ \\ \downarrow \xi \\ (Q, m, e) \end{array}$$

a unique homomorphism ξ with $\xi \cdot \eta Q = \mathrm{id}_Q$. The facts that $m(q, q') = \xi(qq')$, considering qq' as an element of $Q^* = TQ$, while $e = \xi(\Lambda)$ proves that ξ completely determines the monoid structure. In fact, if (Q', m', e') is another monoid – inducing $\xi' : TQ' \longrightarrow Q'$ – and if $f: Q \longrightarrow Q'$ is a function, then $f: (Q, m, e) \longrightarrow (Q', m', e')$ is a monoid homomorphism if and only if

2
$$\begin{array}{ccc} TQ & \xrightarrow{Tf} & TQ' \\ \xi \downarrow & & \downarrow \xi' \\ Q & \xrightarrow[f]{} & Q' \end{array}$$

To prove this we must first recall (7.2.9) that Tf is the unique monoid homomorphism extending f, i.e.

3
$$\begin{array}{ccc} Q & \xrightarrow{\eta Q} & TQ \\ f \downarrow & & \downarrow Tf = UFf \\ Q' & \xrightarrow[\eta Q']{} & TQ' \end{array} \qquad \begin{array}{c} FQ \\ \downarrow Ff \\ FQ' \end{array}$$

Thus, if f is a monoid homomorphism then, since ξ, Tf and ξ' are monoid homomorphisms and since

$$\xi' \cdot Tf \cdot \eta Q = \xi' \cdot \eta Q' \cdot f = f = f \cdot \xi \cdot \eta Q,$$

2 commutes. Conversely, if 2 commutes, evaluating on Λ proves $f(e) = e'$ whereas evaluating at qq' proves $f(q \cdot q') = f(q) \cdot f(q')$.

This observation – that monoid homomorphisms are characterized by 2 – allows us to rewrite 1. First, let $\mu Q : TFQ \longrightarrow TQ$ denote the ξ induced by the monoid FG. Then 1 – the definition of ξ for arbitrary (Q, m, e) – is completely captured by

4
$$\begin{array}{ccc} Q & \xrightarrow{\eta Q} & TQ \\ & \searrow^{\mathrm{id}_Q} & \downarrow \xi \\ & & Q \end{array} \qquad \begin{array}{ccc} TTQ & \xrightarrow{T\xi} & TQ \\ \mu Q \downarrow & & \downarrow \xi \\ TQ & \xrightarrow[\xi]{} & Q \end{array}$$

which is just the **T**-algebra diagram of 2.9! Let us see that this (T, η, μ) is indeed a monad: Since $Tf: FQ \longrightarrow FQ'$ is a monoid homomorphism we have

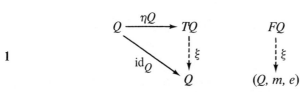

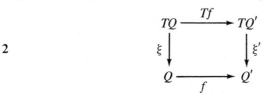

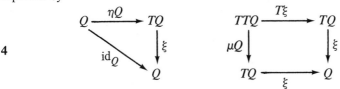

5
$$\begin{array}{ccc} TTQ & \xrightarrow{TTf} & TTQ' \\ {\scriptstyle \mu Q}\downarrow & & \downarrow{\scriptstyle \mu Q'} \\ TQ & \xrightarrow[Tf]{} & TQ' \end{array}$$

i.e. $\mu : TT \dot{\rightarrow} T$ is a natural transformation. Setting $\xi = \mu Q$ in 4 we have

6
$$\begin{array}{ccc} TQ & \xrightarrow{\eta TQ} & TTQ \\ & {\scriptstyle \mathrm{id}_Q}\searrow & \downarrow{\scriptstyle \mu Q} \\ & & TQ \end{array} \qquad \begin{array}{ccc} TTTQ & \xrightarrow{T\mu Q} & TTQ \\ {\scriptstyle \mu TQ}\downarrow & & \downarrow{\scriptstyle \mu Q} \\ TTQ & \xrightarrow[\mu Q]{} & TQ \end{array}$$

Thus we have almost proved that $\mathbf{T} = (T, \eta, \mu)$ is a monad. The only missing detail is

7
$$\begin{array}{ccc} TQ & \xrightarrow{T\eta Q} & TTQ \\ & {\scriptstyle \mathrm{id}}\searrow & \downarrow{\scriptstyle \mu Q} \\ & & TQ \end{array}$$

But since $T\eta Q$, μQ are monoid homomorphisms and, from the naturality square

$$\begin{array}{ccc} Q & \xrightarrow{\eta Q} & TQ \\ {\scriptstyle \eta Q}\downarrow & & \downarrow{\scriptstyle \eta TQ} \\ TQ & \xrightarrow[T\eta Q]{} & TTQ \end{array}$$

we have

$$\mu Q \cdot T\eta Q \cdot \eta Q = \mu Q \cdot \eta TQ \cdot \eta Q = \mathrm{id}_{TQ} \cdot \eta Q.$$

Therefore, 7 commutes.

The discussion thus far may be summarized as follows. $\mathbf{T} = (T, \eta, \mu)$ is a monad and via the passage $(Q, m, e) \leftrightarrow (Q, \xi)$ as in 1, **Mon** is a full subcategory of the category of **T**-algebras with morphisms as in 2.

The one remaining question is whether an ***arbitrary*** **T**-algebra (Q, ξ) comes from a monoid for the **T** defined in this way from $TQ = Q^*$. To see that this is indeed so, define

$$m(q, q') = \xi(qq'), \qquad e = \xi(\Lambda).$$

Recall that

$$(Tf)(q_1 \ldots q_n) = f(q_1) \ldots f(q_n)$$

and that

$$\eta Q(q) = q \text{ (}\textit{qua}\text{ word of length one).}$$

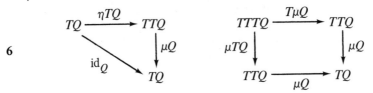

To show that m is associative we use both diagrams of 4: Let $q, q', q'' \in Q$. Consider the word $(qq')(q'') \in TTQ$ (of length two). Then

$$(\xi \cdot \mu Q)[(qq')(q'')] = \xi(qq'q'')$$

whereas

$$\xi \cdot T\xi[(qq')(q'')] = \xi(\xi(q, q')\xi(q'')) = \xi(\xi(q, q'), q'') \qquad \text{(because } \xi \cdot \eta Q = \text{id).}$$

Thus

$$m(m(q, q'), q'') = \xi(qq'q'').$$

Similarly, starting with $(q)(q'q'')$,

$$m(q, m(q', q'')) = \xi(qq'q'').$$

Thus, m is associative. Using both diagrams of 4 again,

$$m(q, e) = \xi(q\xi(\Lambda)) = \xi(\xi(q)\xi(\Lambda)) = (\xi \cdot T\xi)((q)(\Lambda)) = (\xi \cdot \mu Q)((q)(\Lambda)) = \xi(q) = q.$$

Similarly, $m(e, q) = q$.

We have shown, then, that this monad **T** – which is determined by the structure of free monoids – recovers the arbitrary monoids as its algebras. This illustrates the well-known principle of universal algebra: *all information is in the free algebras.*

Diagram 2 begs for the following general definition:

8 **DEFINITION**: Let $\mathbf{T} = (T, \eta, \mu)$ be any monad in $\mathbf{K}$ and let (Q, ξ), (Q', ξ') be **T**-algebras. A morphism $f: Q \longrightarrow Q'$ is a **T-homomorphism** $(Q, \xi) \longrightarrow (Q', \xi')$ just in case

$$\begin{array}{ccc} TQ & \xrightarrow{\;Tf\;} & T'Q' \\ {\scriptstyle \xi}\downarrow & & \downarrow{\scriptstyle \xi'} \\ Q & \xrightarrow[\;f\;]{} & Q' \end{array}$$

Defining composition and identities at the level of **K**, we have the category $\mathbf{K}^{\mathbf{T}}$ of **T**-algebras and **T**-homomorphisms equipped with underlying **K**-object functor

$$U^{\mathbf{T}} : \mathbf{K}^{\mathbf{T}} \longrightarrow \mathbf{K} : (Q, \xi) \mapsto Q.$$

9 **THEOREM**: Let $\mathbf{T} = (T, \eta, \mu)$ be a monad in **K**. Then $U^{\mathbf{T}}: \mathbf{K}^{\mathbf{T}} \longrightarrow \mathbf{K}$ has a canonical left adjoint whose induced monad, as in 2.4, is exactly **T**, i.e. $(TQ, \mu Q)$ equipped with 'insertion of generators' ηQ is the **free T-algebra** over Q.

Proof: Let Q be an object in **K**. Using two of the **T**-laws, 6, $(TQ, \mu Q)$ is a **T**-algebra. Now let (Q', ξ') be a **T**-algebra and let $f: Q \longrightarrow Q'$ be arbitrary.

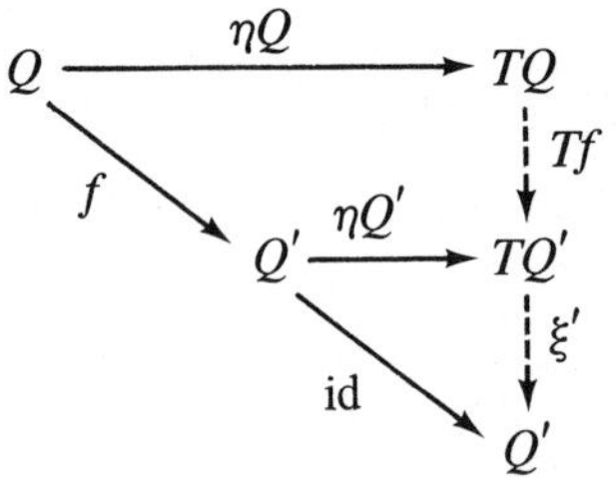

Since $Tf : (TQ, \mu Q) \longrightarrow (TQ', \mu Q')$ is a **T**-homomorphism (μ is natural, see 5) and since $\xi' : (TQ', \mu Q') \longrightarrow (Q', \xi')$ is a **T**-homomorphism (one of the algebra laws!) $\xi' \cdot Tf : (TQ, \mu Q) \longrightarrow (Q', \xi)$ is a **T**-homomorphism which, as shown above, satisfies $(\xi' \cdot Tf) \cdot \eta Q = f$.

Now suppose $\psi : (TQ, \mu Q) \longrightarrow (Q', \xi')$ is a **T**-homomorphism satisfying $\psi \cdot \eta Q = f$. To prove that $\psi = \xi' \cdot Tf$ – and this is where the other law, 7, comes in – glance at

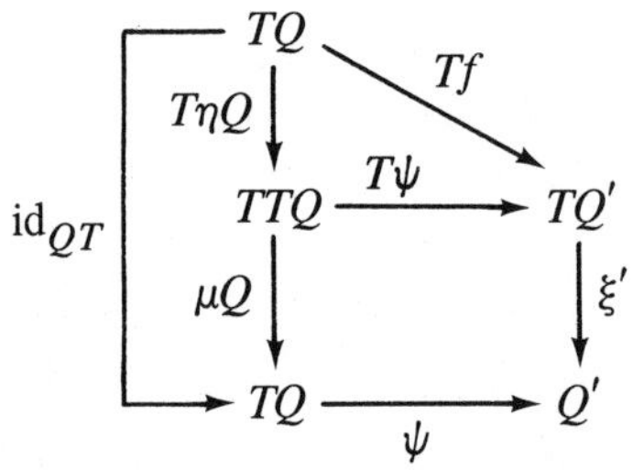

This establishes that the functor $F^{\mathbf{T}} : \mathbf{K} \longrightarrow \mathbf{K}^{\mathbf{T}}$ defined by $F^{\mathbf{T}}(Q) = (TQ, \mu Q)$ and acting on morphisms by $Ff = Tf : (TQ, \mu Q) \longrightarrow (TQ', \mu Q')$ is a left adjoint of $U^{\mathbf{T}}$. It is clear that we need only establish that $\mu : TT \dot{\longrightarrow} T = U\varepsilon F : UFUF \dot{\longrightarrow} UF$. To see this we need only remember that $\varepsilon(Q, \xi) : QF^{\mathbf{T}} \longrightarrow (Q, \xi)$ is defined as the unique **T**-homomorphic extension

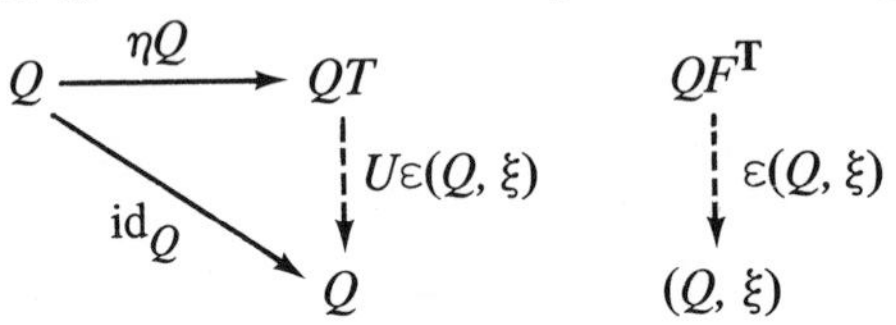

□

We conclude the chapter by clarifying the relationship between algebras of a monad in **Set** and the (Ω, E)-algebras of Section 1. We will provide an outline, referring the reader to the first chapter of "Algebraic Theories" by E. G. Manes (Springer-Verlag, 1975) for a full treatment.

To begin, the inclusion functor (Ω, E)-**Alg** $\longrightarrow$ Ω-**Alg** has a left adjoint. We define

$$\theta : (Q, \delta) \longrightarrow (Q/R, \bar{\delta})$$

for an arbitrary Ω-algebra (Q, δ) by considering the equivalence relation R on Q of all (q_1, q_2) such that for *every* Ω-homomorphism $f: (Q, \delta) \longrightarrow (Q', \delta')$ with (Q', δ') an (Ω, E)-algebra, $f(q_1) = f(q_2)$. The reasoning is that "pairs in R are just the consequences of the equations." Formally, one checks first that there is a unique Ω-algebra structure $\bar{\delta}$ on Q/R admitting θ as an Ω-homomorphism, i.e. that "$\bar{\delta}_\omega(\theta(q_1), \ldots, \theta(q_n)) = \theta\delta_\omega(q_1, \ldots, q_n)$" is well-defined for all ω in Ω_n. Secondly, one must prove $(Q/R, \bar{\delta})$ is an (Ω, E)-algebra. The universal property

$$\begin{array}{ccc} (Q, \delta) & \xrightarrow{\theta} & (Q/R, \bar{\delta}) \\ & \searrow^{f} & \downarrow{\scriptstyle \psi} \\ & & (Q', \delta') \end{array}$$

is then immediate from the definition of R.

Since Ω-**Alg** $\longrightarrow$ **Set** has a left adjoint by 1.12 we have from the composition theorem (exercise 7.3.9) that (Ω, E)-**Alg** $\longrightarrow$ **Set** has a left adjoint; indeed, the underlying set of the free (Ω, E)-algebra on the set Q is obtained as $TQ = (T_\Omega Q)/R$.

For each (Ω, E)-algebra (Q, δ) we may define $\xi: TQ \longrightarrow Q$ as in 1. By the same reasoning as in 2 and 3 we have that every Ω-homomorphism between (Ω, E)-algebras, $f: (Q, \delta) \longrightarrow (Q', \delta')$ induces the commutative square

10
$$\begin{array}{ccc} TQ & \xrightarrow{Tf} & TQ' \\ {\scriptstyle \xi}\downarrow & & \downarrow{\scriptstyle \xi'} \\ Q & \xrightarrow[f]{} & Q' \end{array}$$

The proof that "if f satisfies 10 then f is an Ω-homomorphism" proceeds along the following lines. Let $\omega \in \Omega_n$. Then choose a convenient n-element set $V_n = \{v_1, \ldots, v_n\}$ and consider the element $\omega[v_1, \ldots, v_n]$ of $T_\Omega(V_n)$ and its image $\bar{\omega} = \theta(\omega[v_1, \ldots, v_n])$ in TV_n. By the Yoneda Lemma 7.3.13, $\bar{\omega}$ induces a natural transformation α_ω

$$\alpha_\omega Q : Q^n \longrightarrow TQ, \quad n \xrightarrow{f} Q \mapsto (Tg)\bar{\omega}$$

where $g: V_n \longrightarrow Q$ is defined by $g(V_i) = f(i)$. The essential computation is to verify that $\delta_\omega : Q^n \longrightarrow Q = \xi \cdot \alpha_\omega Q$. For then

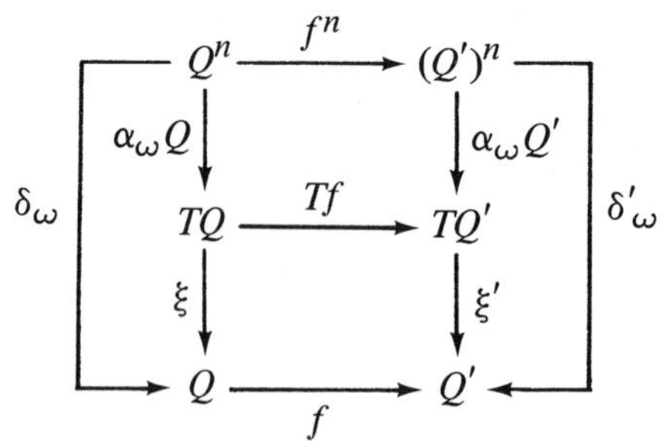

proves that f is an Ω-homomorphism. (The reader should chase this construction for monoids.)

The above construction also explains how to discover the (Ω, E)-algebra corresponding to the arbitrary **T**-algebra (Q, ξ). *Define* $\delta_\omega = \xi \cdot \alpha_\omega Q$. Thus, it is possible to prove the

11 **THEOREM**: The forgetful functor (Ω, E)-**Alg** $\longrightarrow$ **Set** has a left adjoint. (Ω, E)-**Alg** is isomorphic as a category of sets with structure to the algebras of the monad induced by the adjointness as in 2.4. □

A monad $\mathbf{T} = (T, \eta, \mu)$ in **Set** is **finitary** if for every set Q and for every $p \in TQ$ there exists a finite subset F of Q with inclusion $i : F \longrightarrow Q$ such that p is in the image of Ti. In other words, "every term has at most finitely many variables." It is possible to prove that if **T** is finitary then $\mathbf{Set}^{\mathbf{T}}$ is isomorphic as a category of sets with structure to (Ω, E)-**Alg** for some (Ω, E). Thus

12 "Finitary (Ω, E)-Algebras" as defined in Section 1 is coextensive with finitary monads in **Set** insofar as either can be used to define structure on sets.

With this, our indication of the suitability of category theory as a setting for universal algebra is complete. For a wealth of further information on this topic, we refer the reader to the book "Algebraic Theories" cited above.

Exercises

1 Show that the monad of exercise 2.3 is not finitary.

2 Verify directly that the monad for **Mon**, $TQ = Q^*$, is finitary.

3 Show that, in the context of exercise 2.3, the **T**-homomorphisms are precisely the supremum-preserving maps.

4 We say a subcategory **K** of a category **L** is **reflexive** when the inclusion functor $\mathbf{K} \longrightarrow \mathbf{L}$ has a left adjoint $\theta : \mathbf{L} \longrightarrow \mathbf{K}$. We then call θ a **reflection**. Construct and study the reflection $\theta : (Q, \delta) \longrightarrow (Q/R, \bar{\delta})$ when Ω has one binary operation and one nullary operation and (Ω, E)-**Alg** = **Mon**.

5 Construct free groups and free abelian groups using the theory of this section.

6 Relate the present theory to the study of **Dyn**(X) in Section 6.3.

SUGGESTIONS FOR FURTHER READING

Abstract algebra and topology are cornerstones of twentieth century pure mathematics. Abstract groups were introduced by Galois in the early 1800's. Vector spaces and monoids came later. Metric spaces were created in the doctoral dissertation of M. Frechet in 1906, and were generalized to topological spaces soon thereafter.

Categories were first defined by S. Eilenberg and S. Mac Lane in their founding paper "General Theory of Natural Equivalences" in Trans. Amer. Math. Soc. **58**, 1945, 231-294. The motivation was to make precise the concept of a natural transformation which is necessary, for example, to justify why a finite-dimensional vector space is *naturally* isomorphic to its second dual, but only 'unnaturally' isomorphic to its first dual.

We now give a very selective set of suggestions for further reading – many of the references have extensive bibliographies.

Books on Category Theory

H. Herrlich and G. E. Strecker, *Category Theory*, Allyn and Bacon, 1973; 400 pages. In scope, this book deals with the ideas of our first eight chapters, but in much more detail. The reader is offered examples from algebra, topology and analysis (without background treatment). A very complete bibliography is provided.

S. Mac Lane, *Categories for the Working Mathematician*, Springer-Verlag, 1972; 262 pages. An uncluttered introduction, for pure mathematicians, to most of the important ideas of category theory, by one of its founders. As we mentioned in the preface, *Imperative* has as one of its aims to make Mac Lane's book accessible to a wider audience. Historical notes are provided.

A Specturm of Books about Algebra

P. Cohn, *Universal Algebra*, Harper and Row, 1965; 321 pages.

S. Hu, *Elements of Modern Algebra*, Holden-Day, 1965; 208 pages.

S. Mac Lane and G. Birkhoff, *Algebra*, MacMillan Company, 1967; 598 pages.

E. G. Manes, *Algebraic Theories*, Springer-Verlag, 1975.

Elementary Introductions to Topological and Metric Spaces

D. Bushaw, *Elements of General Topology*, John Wiley, 1963; 166 pages. Contains a historical introduction.

J. Greever, *Theory and Examples of Point-Set Topology*, Wadsworth, 1967; 130 pages.

G. McCarty, *Topology: an Introduction with Application to Topological Groups*, McGraw-Hill, 1967; 270 pages.

B. Mendelson, *Introduction to Topology*, Allyn and Bacon, 1968; 202 pages.

Applications to Computation and Control

L. S. Bobrow and M. A. Arbib, *Discrete Mathematics: Applied Algebra for Computer and Information Science*, W. B. Saunders, 1974; 719 pages. We recommend the final chapter "Machines in a Category" as a sequel to the brief glimpses in *Imperative*.

E. G. Manes (Ed.), *Category Theory Applied to Computation and Control*, Springer Lecture Notes in Computer Science. Includes background articles, research papers by many contributors, and an extensive bibliography.

INDEX

INDEX